Contents

1. Introduction
2. Article 156: Sun simulator behind a cloud
3. Article 157: Sun simulator: the reasons for its use
4. Article 211: Planet X and Sun interaction: Sun goes dark on April 18th 2018
5. Article 150: Simulated Blood Red Moon
6. Article 151: The Blood Red Moon Simulator at 30 000 feet
7. Article 160: Sun Simulator: Speeds and Orbits
8. Article 164: Secret advanced technology is being used in our skies
9. Article 165: Sun Simulator: irrefutable evidence
10. Article 123: The biggest scientific cover up: our Sun
11. Article 166: Sun Simulator and lens system
12. Article 213: Producing lenses out of air in the earth's atmosphere
13. Article 214: Global Sun Simulation System and the Dying Sun
14. Article 204: Harmful UVC radiation reaching earth's surface indicates source within atmosphere
15. Article 147: The chemtrail and iridescent cloud connection
16. Article 32: The purpose and effects of chemtrails
17. Article 148: The purpose and effects of chemtrails
18. Article 33: Artificial weather
19. Article 137: Noctilucent clouds, rocket launches and chemtrails: what are they hiding?

Chapter 1
Introduction

It was not until around the middle of 2016 that I noticed that something was not right with our planet and the Solar System. I would take my daily walks close to sunset and notice that clouds would often look pink for some time as the sun set. I knew that the pink illumination could not possibly be coming from the Sun and the pink could be observed on clouds which were above my head and thus very far from the setting sun and the horizon. This led me to investigate what was going on and I quickly discovered that the Solar System had been invaded by objects, which could be observed in the Sun's corona and were obviously drawing energy from the Sun.

I also found that the Sun was going dark at times and that becoming increasingly weak as a result of the presence of these objects which turned out to be a system of dead stars, and which I ended up referring to as Stellar Cores. I also quickly realized that in order for these objects not to be easily seen from the earth's surface and for the fact that the Sun was periodically going completely dark, a system had to be in operation on planet earth, to hide these events from the earth's population.

In this book, I have put together the most important articles I have written detailing both the evidence for what is happening with the Sun and detailing the type of devices and systems that seem to be in use for the purpose of hiding what is going on in our Solar System from the earth's population. Since the spraying of aerosols in the earth's atmosphere or chemtrails also plays a big part in the technology used to hide what is going on in the Solar System, I have also included a few articles I have written on this subject.

Dr Claudia Albers

Planet X physicist

Chapter 2
Article 156: Sun simulator behind a cloud

The photograph below shows what at first may appear to be the Sun behind a cloud. The light from the Sun seems to form dark and light rays. These are produced by shadows resulting from light being interrupted by the cloud. Notice that the rays seem to originate behind the cloud. This is because the source of the light is just behind the cloud. Real sunlight coming from the Sun is made out of rays that are parallel to each other because the Sun is 93 million miles away. The fact that light coming from a light source, at infinity, is made of rays that are parallel to each other is a basic physical principle. The source of the light in this case is close, not at infinity. If we follow the rays and imagine where they intersect, we will find the source of this light, it will not be too far behind the cloud. This light source cannot therefore be the Sun, it is a Sun simulator.

Figure 1.1: A photograph of the ocean and sky reveals that the light source is not far behind the clouds producing the diverging ray effect. In addition, this light source is white not yellow like the real Sun would be. The pink haze in the sky, above the horizon, reveals the presence of extra natural light sources, extra stars, in the Solar System.

In addition, the Sun produces white light and will therefore look white outside the earth's atmosphere, but inside the earth's atmosphere, it looks yellow. The reason for this is that white light is made out of 3 main different wavelengths, red, green and blue. The earth's atmosphere scatters the blue wavelength, and thus removes it from light coming directly from the Sun. This is why the sky looks blue; the scattered light seems to come from every direction in the sky. But the white light coming directly from the Sun. and thus from the point where we see the Sun in the sky, will have the blue wavelength missing and will therefore look yellow. The fact that the light source in the sky looks white and not yellow tells us that it is not the real sun. This is a white sun simulator in the sky.

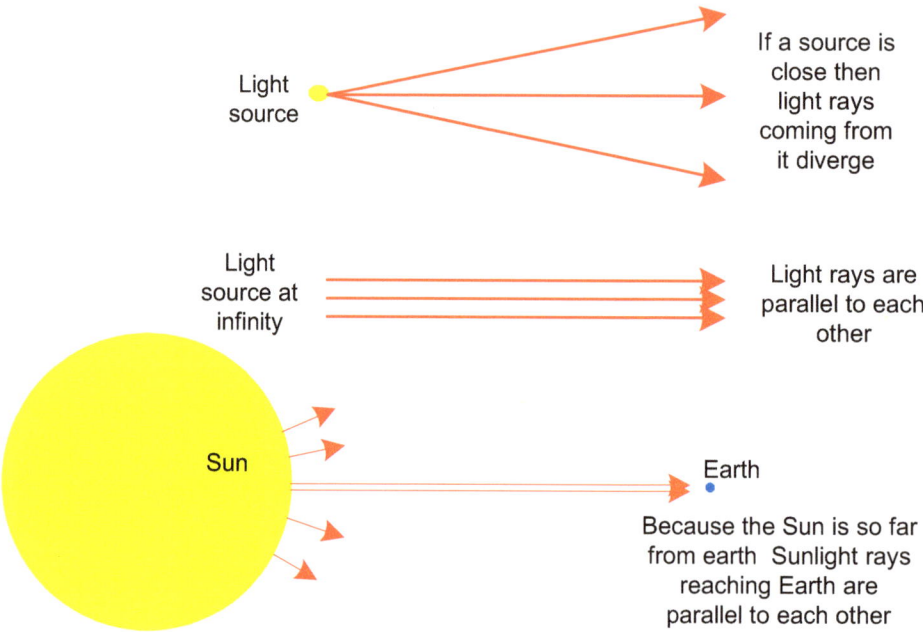

Figure 1.2. Illustration of the basic physics principle which states that light coming from a source at infinity must be parallel to each other. This shows that the source of light behind the cloud in figure 1 cannot be the real sun.

Figure 1.3. Combining the 3 different primary colors of light produces other colors, cyan from combining blue and green, yellow from combining red and green, and **magenta** from combining red and blue. When all 3 primary colors are combined equally, white light is produced.

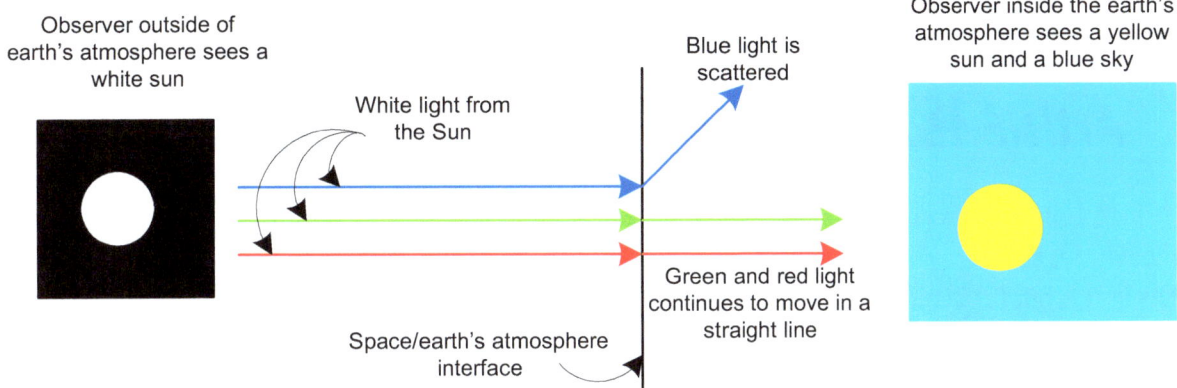

Figure 1.4. White light from the Sun is made up of red, green and blue. Blue is scattered by the earth's atmosphere so only green and red continue in a straight line. Green and red combined produces yellow light so the Sun looks white, outside the atmosphere, and yellow, inside it. The sky looks blue.

The next interesting observation to be made, about the photograph in figure 1, is the pinky orange light, illuminating the clouds, in the photograph, and which also appears in the sky, above the horizon. This color is not natural to our Sun, or our planet's atmosphere, and is evidence of extra stars, which give off this color of light, and have positions close to the Sun's position, and that are illuminating the earth's atmosphere with these colors.

In conclusion, the white light source in the photograph looking out over the ocean, shown in figure 1, showing rays of light, diverging from behind the cloud, is a Sun simulator, not the natural Sun. The Sun simulator's position is at the point where the rays of light intersect and thus not far behind the cloud. The light source is therefore inside the earth's atmosphere.

Chapter 3

Article 157: Sun Simulator: The reasons for its use

In Article 156: Sun simulator behind a cloud [1], I showed that a Sun simulator is operating in the Earth's atmosphere, as the rays produced as a result of light being interrupted by a cloud diverge away from a point that is not far behind the cloud, or far off the surface of the ocean, which means that the source of the light is in the earth's atmosphere, and cannot be the real Sun. The real Sun is very far away, or at infinity, and it is a basic physical principle that light rays, coming from infinity, are parallel to each other, so the real sun cannot produce diverging rays, as we see in the photograph below. In this article, I will focus on the reason why this simulator is being used, and in answering several questions that usually crop up regarding Sun simulators.

Figure 2.1: A photograph of the ocean and sky reveals that the light source is not far behind the clouds producing the diverging ray effect. In addition, this light source is white, not yellow, like the real Sun would be. The pink haze in the sky, above the horizon, reveals the presence of extra natural light sources, extra stars, in the Solar System.

In Article 118: Solar activity declining independent of the solar cycle: is the Sun dying? [2] I showed that the Sun has been weakening, for a long time, and the weakening effect seems to have accelerated in 2016 and 2017. Some of the evidence for this is shown in figures 2 and 3 below. The Sun is thus expected to continue to weaken, at an accelerated rate, which should lead to its light emission to diminish further.

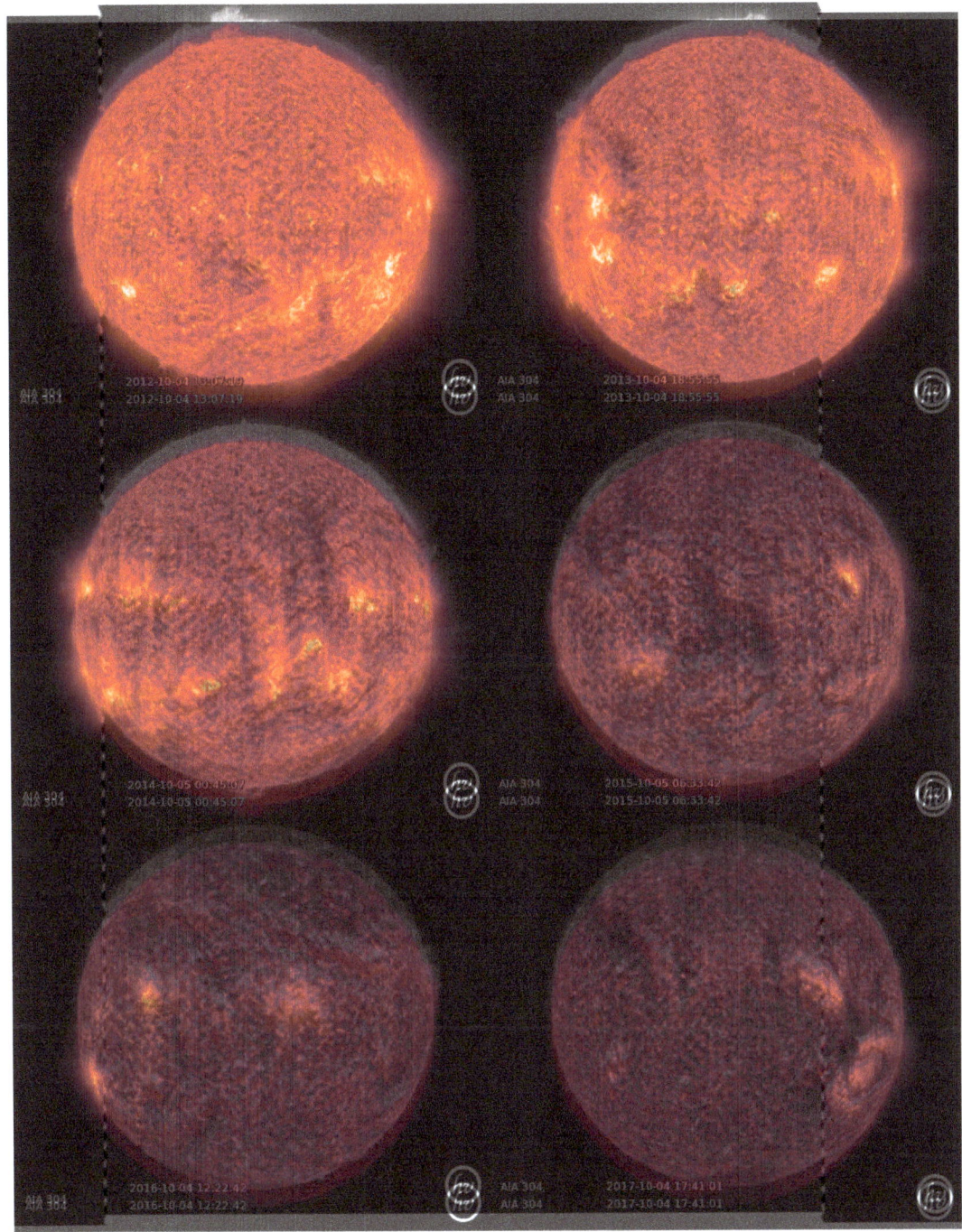

Figure 2.2: SDO images of the Sun in the 304 angstrom wavelength from 2012, 2013, 2014, 2015, 2016 and 2017. The Sun has clearly grown darker, in this wavelength, over the years. Comparison of the brightest spots, in the different images, indicates that they can achieve the same lighter color and that therefore the increased darkness is not due to a change in the color assigned to different intensity.

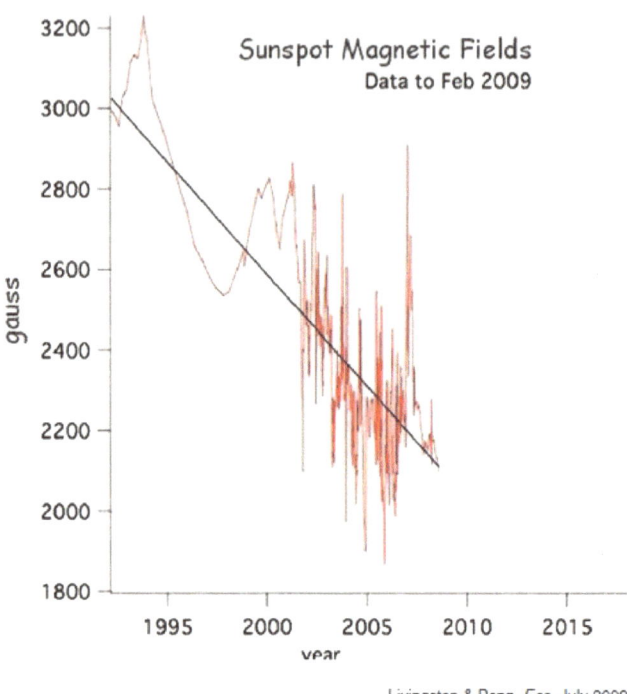

Figure 2.3. The Sun's magnetic field strength, associated with sunspots, dropped independent of the solar cycle during cycle 23. Livingston and Penn found that it consistently dropped by 50 gauss per year between 1996 and 2009 [3].

The reason why this is happening is the presence of the system of old stars, which I usually refer to as Stellar Cores, which have invaded the Solar System and are draining the Sun of energy. Evidence for the presence of these old stars is provided in Article 116: Planet X Objects: unbelievable evidence and size [4]. A few examples, of the observational evidence, for the presence of these objects, in the Sun's corona, is provided below.

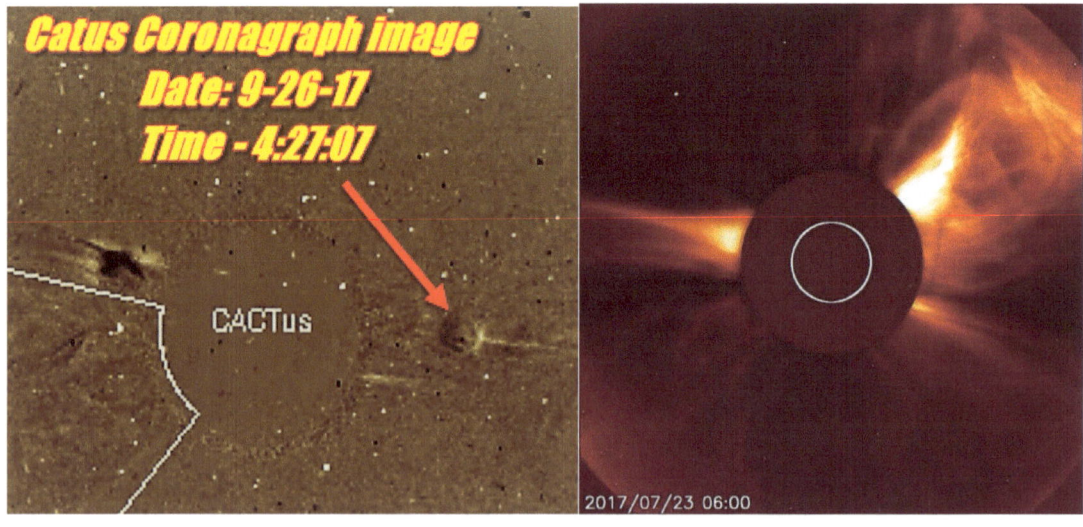

Figure 2.4. Coronagraph images showing Stellar Cores in the Sun's corona during CMEs.

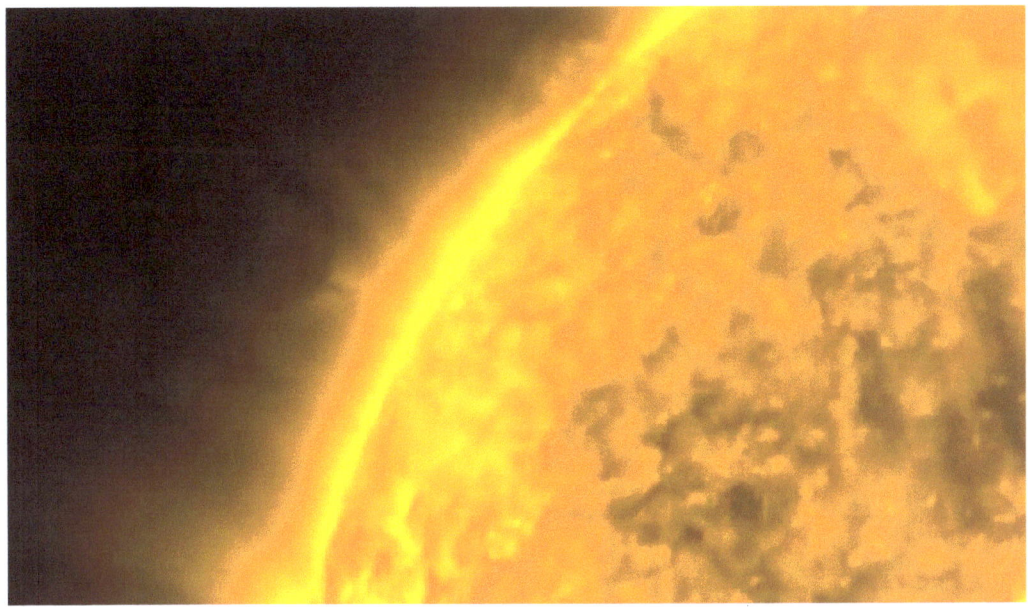

Figure 2.5: SDO image in the 171 angstrom wavelength from October 13th 2017 showing a dark Stellar Core, which appears to be about half of the size of Jupiter. The fact that it is dark in this wavelength may be an indication that this particular Stellar Core is a newer arrival at the Sun's corona.

The fact that the Stellar Cores in the Sun's corona are draining the Sun. and causing light emission from the sun's ionization layers, Article 211: Planet X and the Sun interaction: Sun goes dark on April 18th 2018 [5]. This article appears in the next chapter.

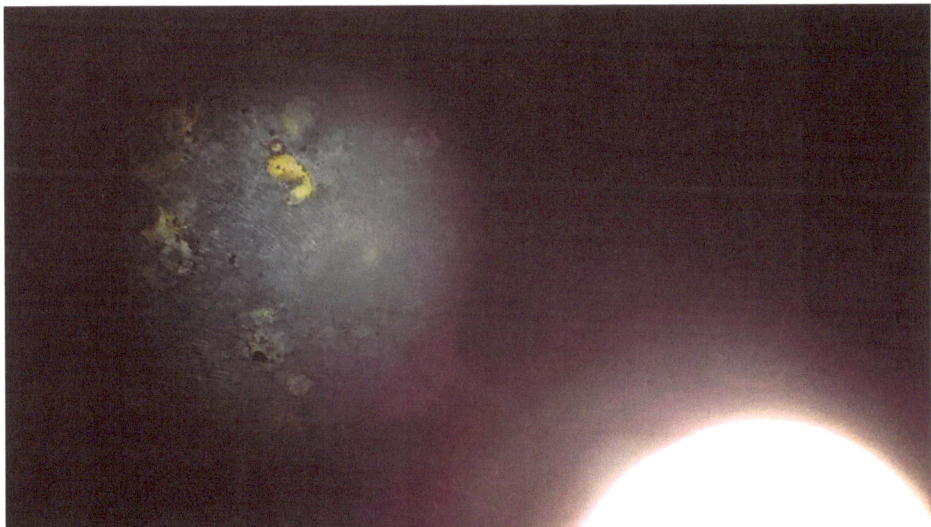

Figure 2.6. Telescopic image of the Blue Stellar Core in the Sun's corona exchanging gaseous plasma with the Sun. The plasma collecting on the Sun's near surface to the object is emitting magenta (pink) light.

Now, the Stellar Cores do not seem to emit visible light, when they arrive at the Sun, but in the process of time, as they absorb energy, and plasma, from the Sun, they regain the ability to emit visible light, in a process, referred to as rejuvenation. The colors of light that these objects emit are not the same as the Sun's, which emits white light. These rejuvenated Stellar Cores seem to emit light of different colors, such as magenta colored light. The Blue Stellar Core, which was in the process of rejuvenating, when photographed in the Sun's corona, through a telescope, is seen in the figure above. The object was starting to emit magenta colored light, from the gaseous plasma, that was gathering under it.

The reason for the use of a Sun simulator seems to therefore be: 1. To hide the fact that the Sun is dying and much weakened; and 2. To hide the fact that there are extra stars, rejuvenated Stellar Cores, close to the Sun and illuminating the Earth's atmosphere.

There is evidence that several different Sun simulation models are in use. There are a few in low earth orbit, and the rest seem to be in the Earth's atmosphere. Evidence regarding the existence of these devices comes from some of the patents available as illustrated in figure 7 below. Figure 8 shows the hexagonal lens array and lamps mounted at the back of the device. Some of the ISS footage which shows that a sun simulator is usually seen in these images instead of the real Sun appears in figure 9. Figure 10 to 13 show the Sun simulator design based on the Sun simulator often seen in ISS video footage. This object would be operating outside the earth's atmosphere. Figure 14 shows a possible design for a device that may be used in the earth's atmosphere. Evidence, for the fact that these simulators are also being used to simulate the moon appears in Article 150: Simulated Blood Red Moon [6] and Article 151: The Blood Red Moon Simulator at 30 000 feet [7], which appear in later chapters in this book:

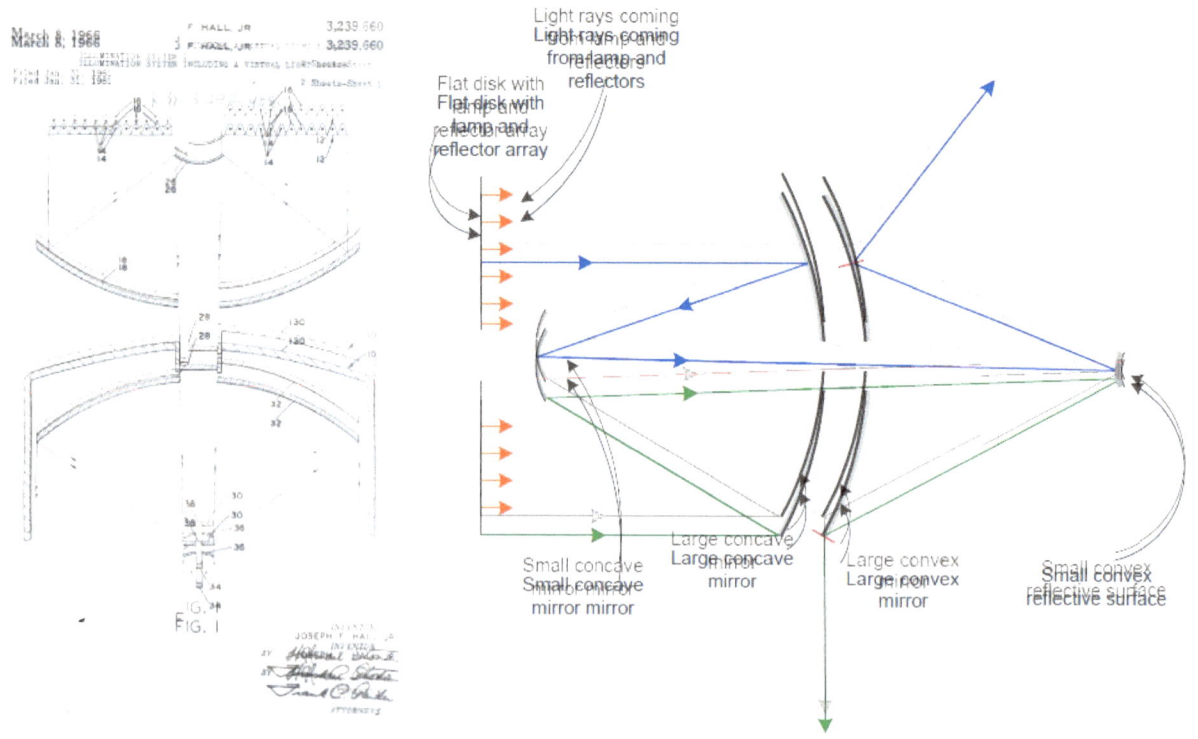

Figure 2.7: Sun simulator diagram from a 1966 patent (left) and my own design diagram based on this patent (right).

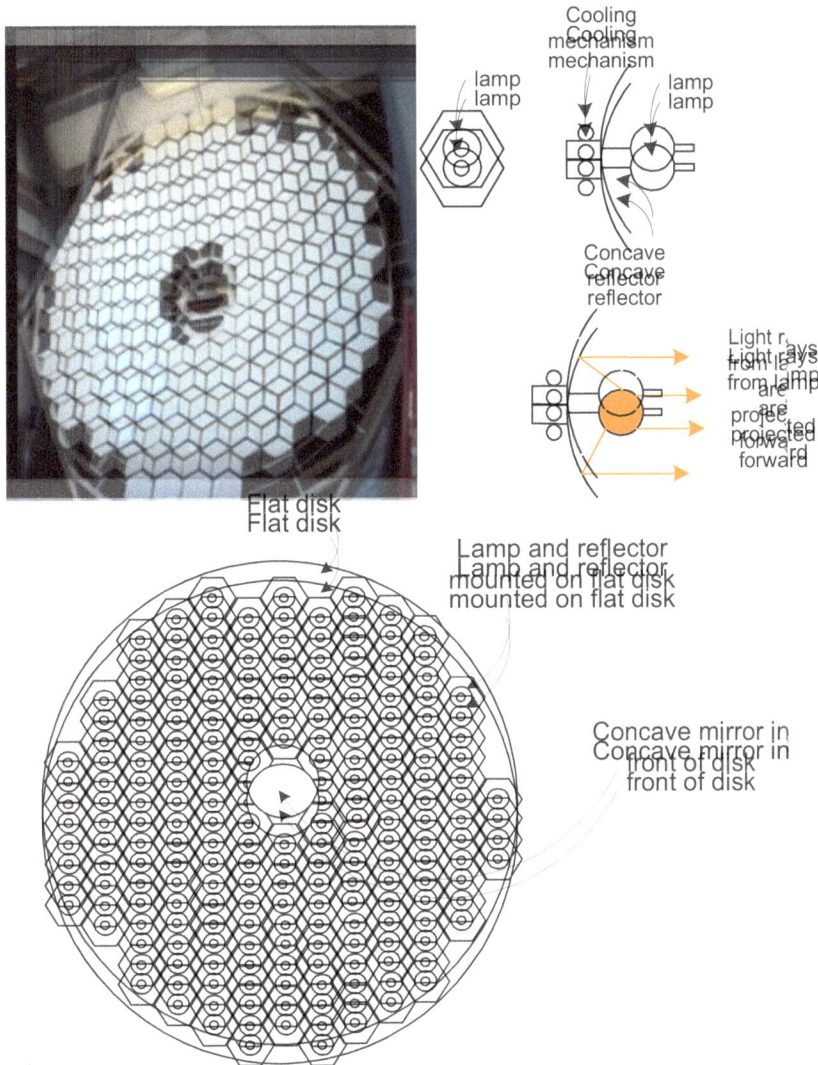

Figure 2.8: Hexagonal lens array, on which hexagonal reflectors and arc lamps, possibly mercury-xenon high pressure arc lamps, are mounted, as well as a cooling mechanism for each lamp.

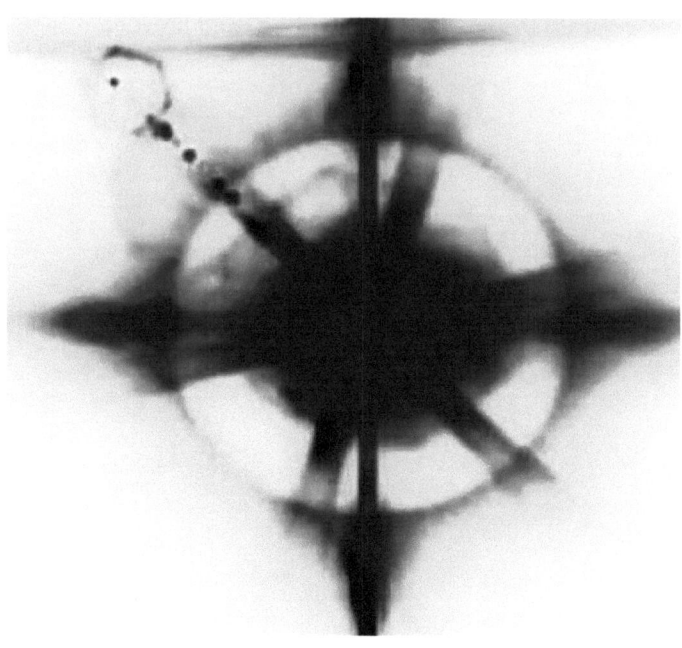

Figure 2.9. Photographs used for the ISS Sun Simulator design diagrams: Top left: Sun simulation device as seen in a video from ISS. Top right: View from earth of the reflector in front of device. Bottom left: Negative image, taken from ISS fotage, of device. Bottom right: Negative image of device simulating the Sun during the August 21st total lunar eclipse (from youtube channel: madsplatta lot).

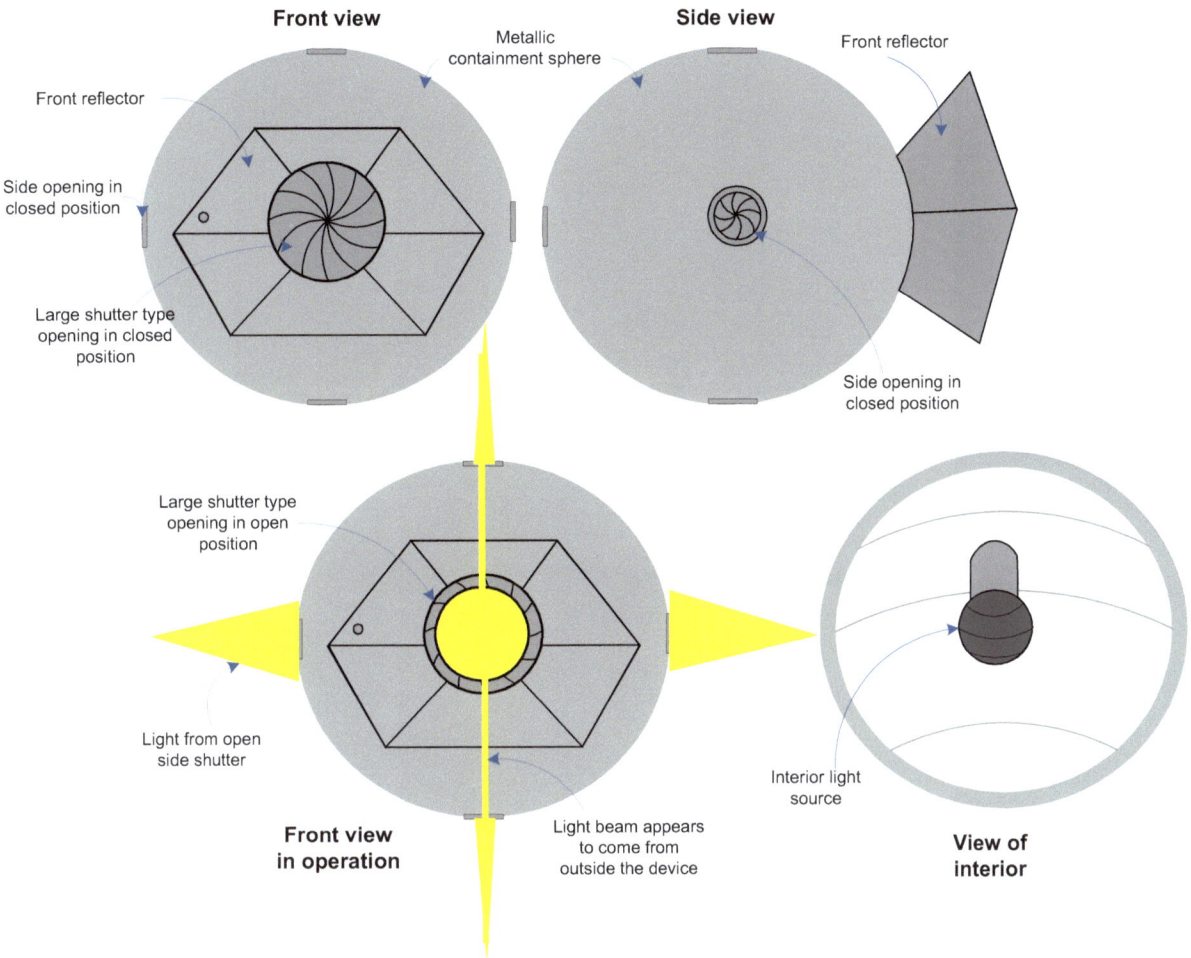

Figure 2.10. Illustration of the design and operation of the ISS Sun simulator: The device has five shutter type portals, and the largest is in front. The front portal is surrounded by an umbrella type of reflector. There are likely to be laser light sources, which reflect off the device, creating the vertical continuous rays of light.

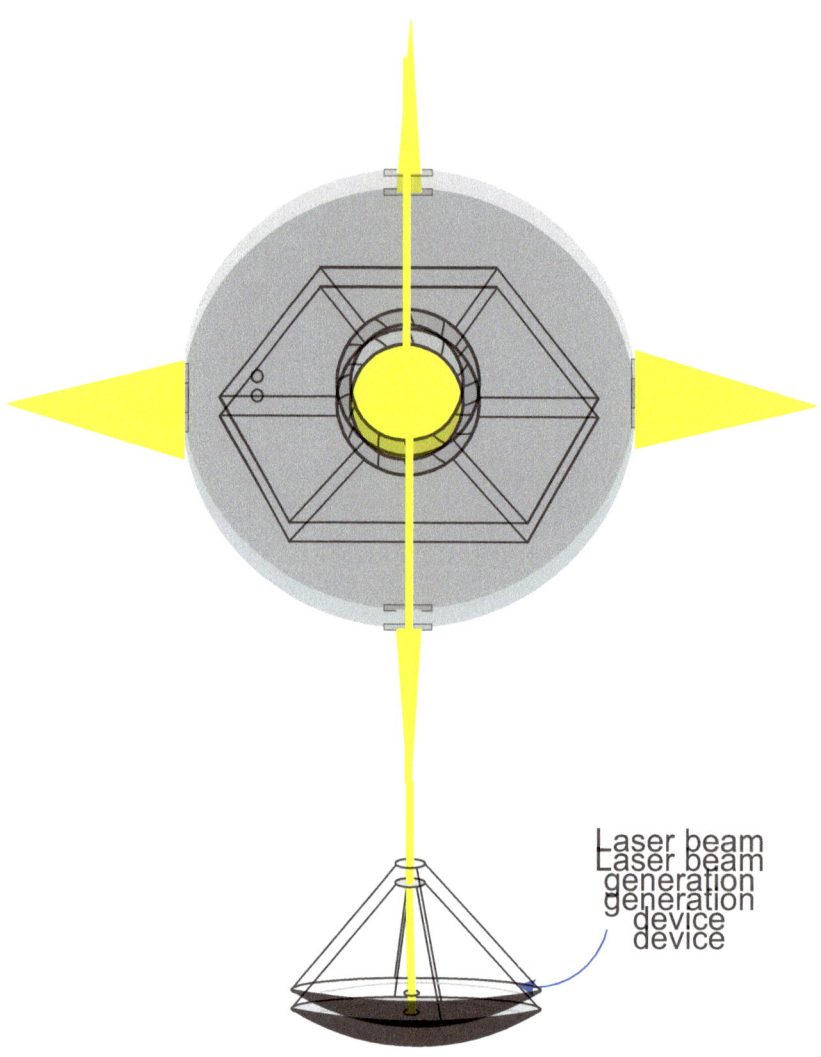

Figure 2.11: The ISS Sun Simulator with a laser beam generation device, below it, giving rise to the vertical light beam, the device is observed to have. Another similar device would be positioned above the device in order to produce the top vertical beam.

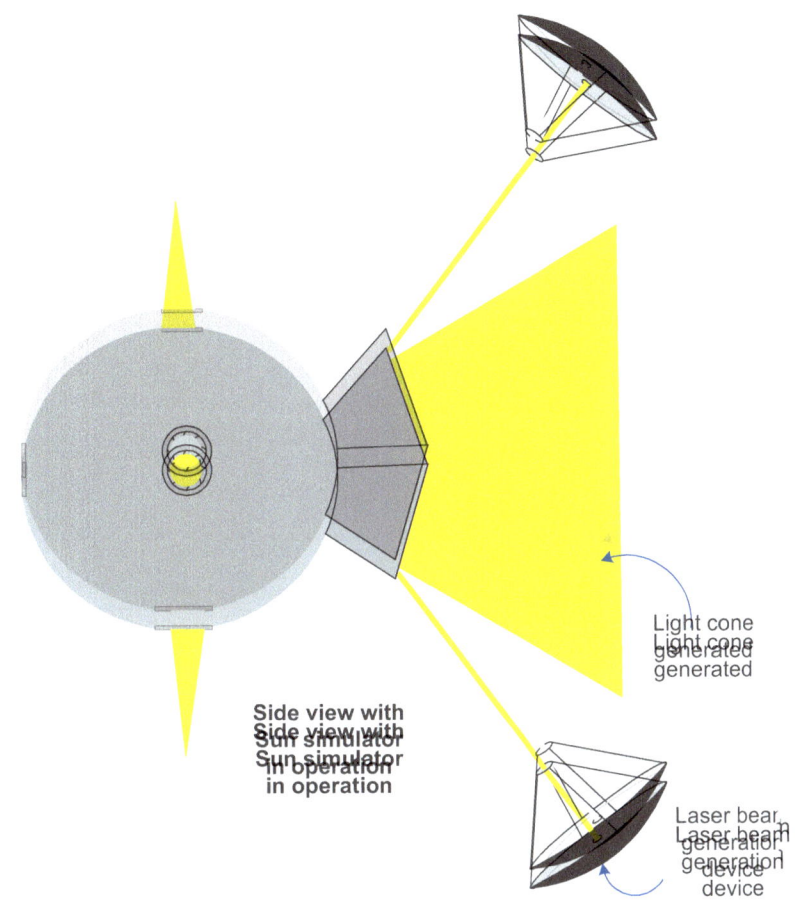

Figure 2.12: The device generates a light cone, which looks circular, like the Sun, from the earth's surface only if the device is directly facing the observer on the surface of the earth. Viewed from the side the cone of light is like the light produced by a flashlight.

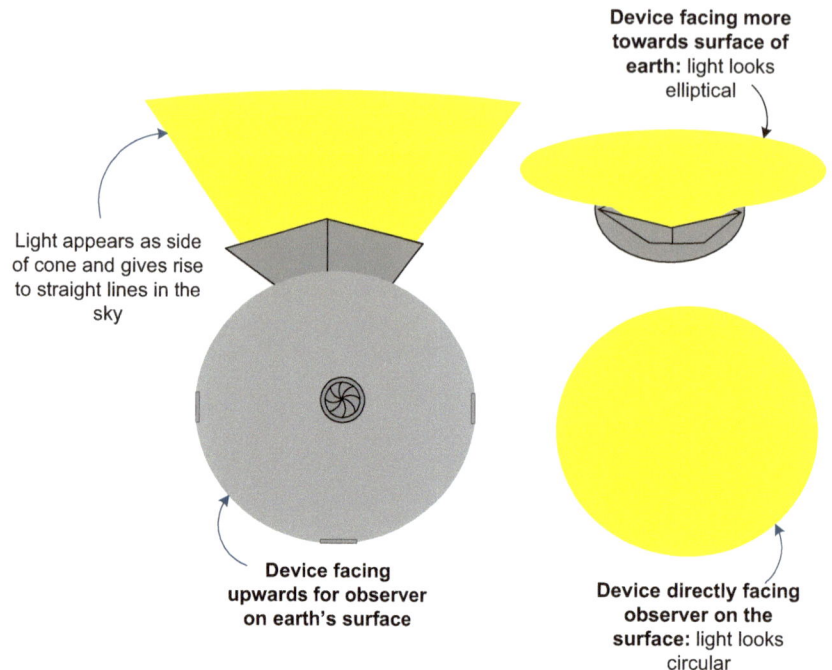

Figure 2.13. The device is able to create sunsets, in which the Sun initially looks elliptical, because it is not fully facing the observer but as it turns toward the observer, the light becomes circular and looks more like the Sun.

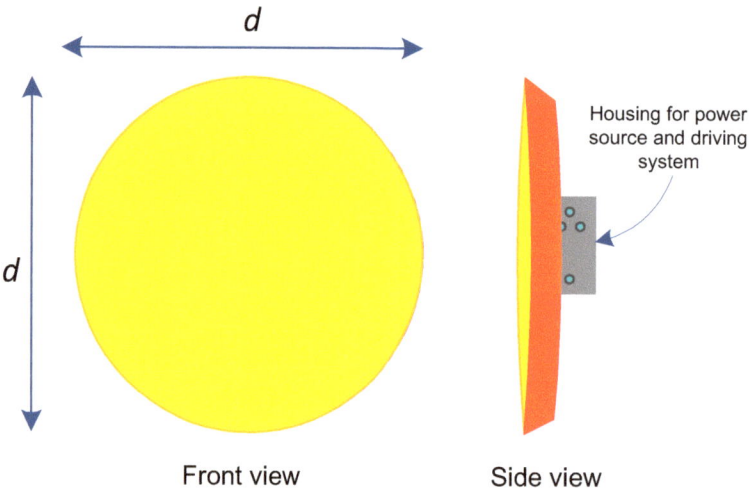

Figure 2.14. Illustration of what a sun/moon simulation device, designed to operate within the earth's atmosphere, may look like. The diameter, d, will vary between 300 ft, for a device designed to operate at 30 000 ft, to 400 ft, for a device designed to operate at 45 000 ft. The device may also have projection capabilities, so that sun and moon features can appear on the light emission surface.

Figure 2.15. Saucer shaped flying craft developed by NAZI scientists before, and during, the Second World War. Their ability to hover silently shows that they used an antigravity propulsion system.

Now, the question often arises as to how these objects stay in the air. The fact is that during the Second World War, the NAZIs had already developed craft, which used antigravity drives that could quietly hover. This technology was absorbed by United States secret research programs. The existence of such technology, and the fact that it has been in existence for 70 years, is detailed in Article 115: The secret space program [8]. Thus, since the technology has been existence for so long, why should it not be used in Sun simulators, which are supposed to remain hidden from the public?

The other question, which has arisen is: wouldn't a simulator, in the atmosphere, cause the Sun to seem larger, than is normal, depending on the distance from the observer? Indeed, it will. The Sun will, and often does look larger, than normal, and people will feel that it is thus especially bright that day and some will complain that it is burning their skin, due to the fact that the simulator is closer to their position than at other times. However, it is difficult to judge, the difference in size, from a light source, with the correct 0.5° angular width, from one with an angular width twice that, in the sky. So the difference usually goes unnoticed and people will just think that the Sun is brighter than normal, at those times.

In conclusion, Sun simulation devices are being used to hide the fact that the Sun is much weaker than normal, continuing to weaken at an accelerating rate, and also to hide the fact that there are extra light sources, in the Solar System. These objects are rejuvenated old stars, or Stellar Cores, which are also responsible for the Sun's weakened state.

References:

[1] Albers, E. (2017). Article 156: Sun simulator behind a cloud.

[2] Albers, E. (2017). Article 118: Solar activity declining independent of the solar cycle: is the Sun dying?

[3] Matthew J. Penn and William Livingston, "Long-term Evolution of Sunspot Magnetic Fields."
http://www.probeinternational.org/Livingston-penn-2010.pdf

[4] Albers, E. (2017). Article 116: Planet X Objects: unbelievable evidence and size.

[5] Albers, E. (2017). Article 211: Planet X and the Sun interaction: Sun goes dark on April 18th 2018.

[6] Albers, E. (2018). Article 150: Simulated Blood Red Moon.

[7] Albers, E. (2018). Article 151: The Blood Red Moon Simulator at 30 000 feet.

[8] Albers, E. (2017). Article 115: The secret space program.

Chapter 3

Article 211: Planet X and Sun interaction: Sun goes dark on April 18th 2018

After noticing that the SDO 193 angstrom 48 hour video loop, available on April 20th 2018, showed evidence of images being partly blocked out, which to me is now a sign that NASA is trying to cover up something going on with the Sun, whilst at the same time allowing a glimpse of what is happening, I looked at the images on Helioviewer.org, and I noticed that all the 193 angstrom images from April 19th at 2:34:41 to April 20th at 00:00:05 (UTC) were missing, whilst in the 211 angstrom, some of the images were available but at least, at 1 hour intervals. The 193 angstrom images between April 18th at 02:15:05 and April 19th at 00:00:05 (UTC) were also missing. But more of the 211 angstrom images were available. This to me was a sign that something major had happened to the Sun on that day and indeed further investigation suggested that the Sun had gone dark on both days. Since the earth's magnetosphere was also impacted by a shock wave, which gave rise to a geomagnetic storm, on April 20th 2018, it is likely that these events are related to each other.

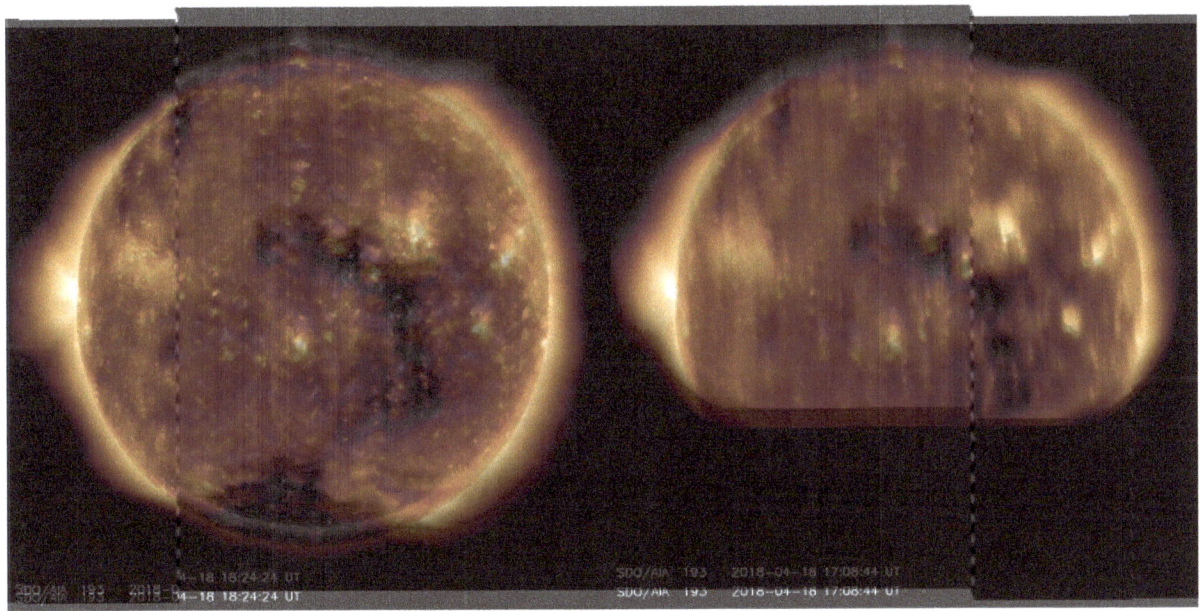

Figure 3.1: SDO images in the 193 angstrom obtained from the SDO 48 hour video loop. The image on the right, from April 18th, shows signs of the Sun going dark in the region cut-off from view as the part of the Sun seen just above the cut-off line is darker then the same region in the left image.

The 211 angstrom SDO images available on Helioviewer.org from April 18th 2018 also indicate that the Sun did indeed go dark on this day.

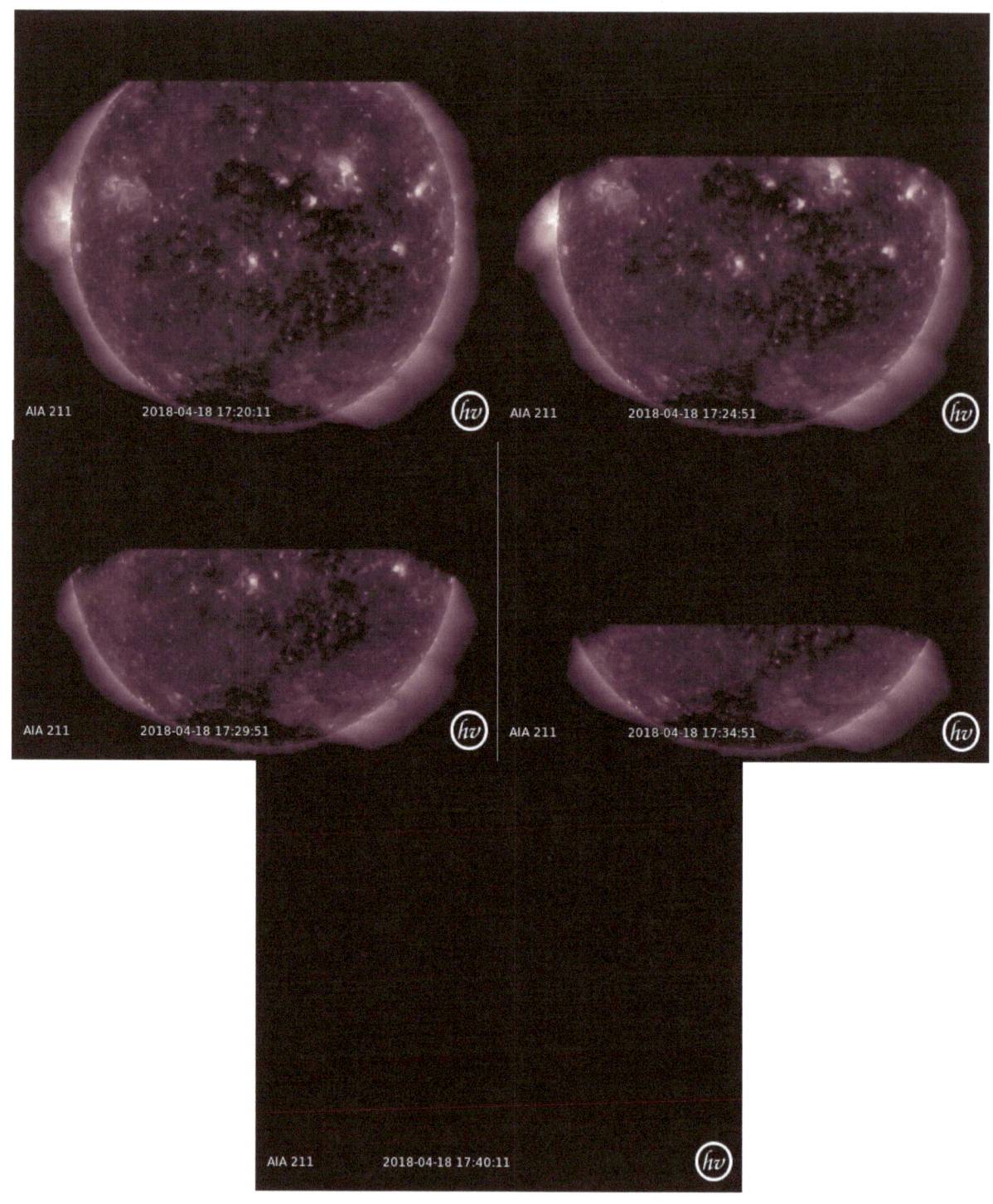

Figure 3.2. Jagged edges, on either side of the Sun, below the cut-off line, indicate that the Sun has gone dark in the region, which has been cut-off from view. The jagged lines are like those that would appear around coronal holes, but previous images, clearly show that there were no coronal holes present, in these regions. The fact that the Sun is not visible at all, in the last image, suggests that the Sun went completely dark at that time.

This is not the first time that the Sun has gone completely dark. It usually goes completely dark during the so called SDO eclipse season. The Sun going dark at these times is supposed to be due to the earth eclipsing the SDO satellite's view of the Sun, but careful examination reveals that the images cannot possibly be due to an eclipse, but that the Sun actually goes completely dark at that time.

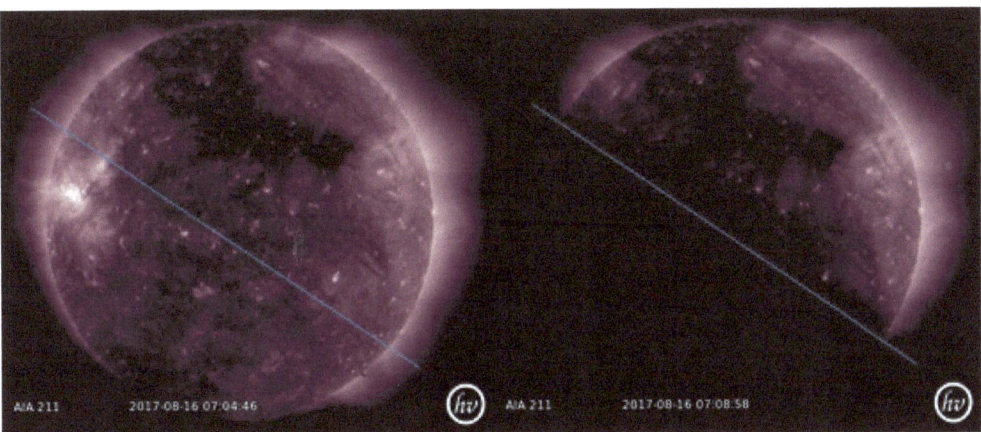

Figure 3.3. SDO images of the Sun from August 16th 2017 (first day of the second eclipse season of 2017) at 7:04 and 7:08 (UTC), in the 21.1 nm wavelength, showing that the Sun's corona shrinks back, instead of being covered by the earth, as we would expect from an eclipse.

Figure 2 compares the 2nd image, in row 1, with the 3rd image, in row 2, and figure 3 below compares the 2nd image in the 1st row with the 2nd image in the 3rd row. The corona clearly shrinks back.

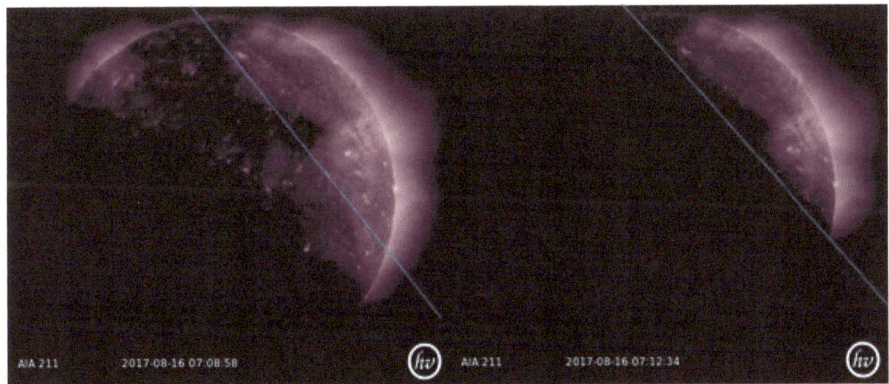

Figure 3.4. SDO images of the Sun from August 16th 2017 at 7:08 and 7:12 (UTC), in the 21.1 nm wavelength, showing that the Sun's corona shrinks further, the shape of the coronal hole close to the blue line changes shape, and the angle, at which the darkness advances, changes (slope of the blue line becomes steeper). None of these are what is to be expected, if the advancing darkness is as a result of the earth eclipsing the Sun (see Article 205: NASA indicates that the Planet X system is affecting the Sun) [1].

The SDO eclipse season appears to be due to an object that is in a regular orbit, around the Sun, which comes by, the Sun, twice per year, the object comes around for its first pass each year, in February, and arrives 4 days earlier each year. There is clear evidence that this has been occurring since 2011 (see Article 110: How do stars produce light?) [2].

The Sun also goes, at least partially, dark, periodically. It went at least partially dark on July 5th 2017 (see Article 26: The Sun goes partially dark on July 5th 2017) [3], and again on October 4th 2017. There is also evidence that the same thing occurred on April 11th 2018.

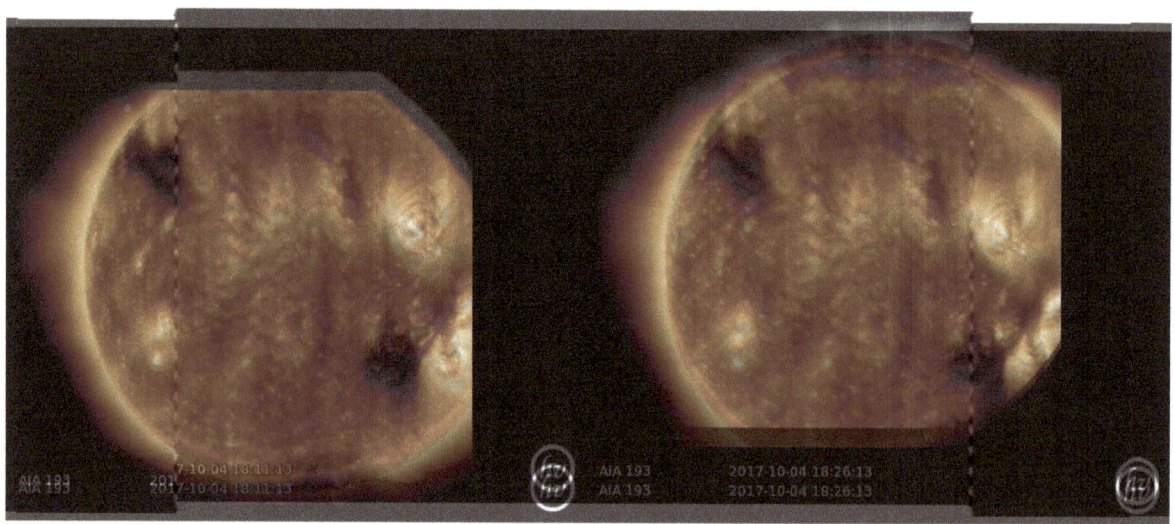

Figure 3.5: SDO images of the Sun in the 193 angstrom wavelength, from October 4th 2017, indicating that the Sun had gone dark in the bottom right hand corner (see Article 60: The Planet X system and the reacting Sun) [4].

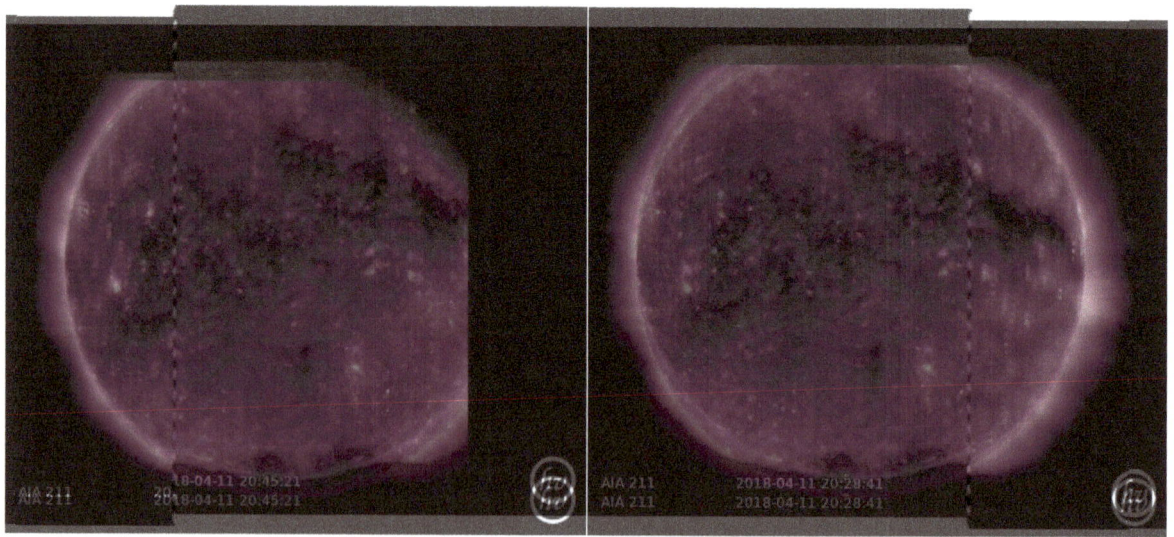

Figure 3.6: SDO images in the 211 angstrom from April 11th 2018 indicating that the Sun had gone dark in the top right hand corner (see Article 205: NASA indicates that the Planet X system is affecting the Sun) [1].

The fact that the April 11th 2018 event was only 7 days from the April 18th 2018 event, suggests that the Sun is going dark more frequently and is thus increasingly more affected by the objects, Stellar Cores, or members of what I refer to as the Planet X System, which are draining the Sun of energy, and causing the Sun to weaken. The weakening trend seems to be accelerating and the Sun will eventually not be able to emit any light (for more details see Article 195: Stellar Cores and the dying Sun) [6].

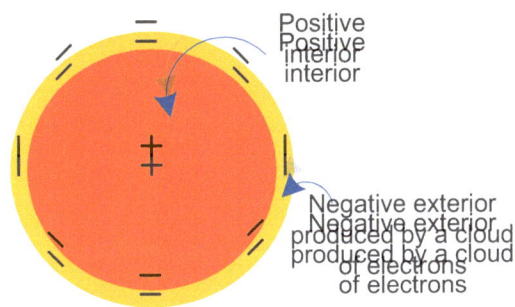

Figure 3.7: All objects in outer space develop a negatively charged outer layer, made of mainly electrons, as a result of the separation of charge part of the gravitational interaction, which causes positively and negatively charged particles to repel each other (see Article 181: Stellar Cores and deciphering gravity [6] and Article 182: Einstein's dream realized: unified field theory of electro gravitation [7] in Book: Planet X Revealed Gravity and Light, for more details).

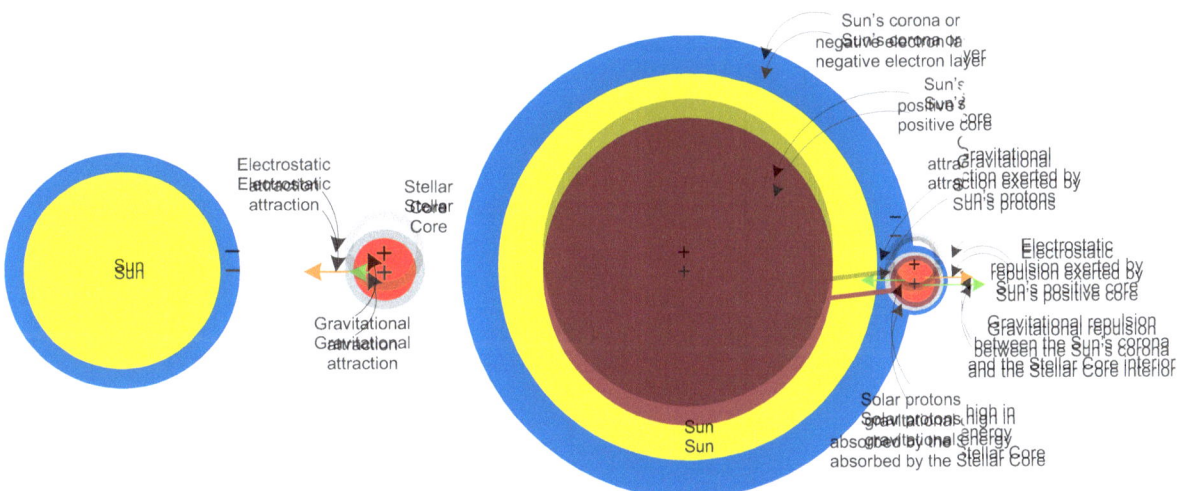

Figure 3.8: Stellar Cores have a neutral outer layer and thus operate like super ions, they are electrostatically attracted to the Sun's negative outer layer. Once in the Sun's corona, they are repelled by the Sun's positive inner core. They draw proton dense matter, or positively charged ions, from the Sun's inner layers. Eventually this causes them to be ejected by the Sun's outer negative layer, as a result of the charge separation gravitational interaction. This causes the Sun to eject matter or a CME (see Article 193: Stellar Cores in the Sun's corona: why do they not collide with the Sun? in the Book: The Light Universe) [8]

As a result of the charge separation part of the gravitational interaction, all objects in the universe have a positively charged core, and a negatively charged outer layer. But Planet X system Stellar Cores are depleted in electrons and thus operate like super ions. They are attracted to the Sun via the electrostatic interaction. When they first make contact with the Sun's corona, which is the Sun's outer negative layer containing mainly electrons, they draw on the Sun's supply of electrons, and in this way cause the Sun's electric potential in its outer layers to drop, which in turn causes the Sun to stop electric discharging (lightning) and thus light emission. Once the dead star has been enveloped in a layer of electrons, it will then start drawing positive ions from the Sun's deeper layers, such as the chromosphere. It will absorb the gravitational energy (photon energy) in the ions it is absorbing, until a point is reached when the gravitational repulsion between the positive ions and the Sun's electron layer reaches a critical level and the Stellar Core is ejected. When this occurs, a large quantity of the matter, it was drawing from the Sun, contained in the matter vortex connecting it to the Sun, is also ejected. This may look like a CME expulsion. The matter will travel across interplanetary space, and if the earth is in its path, it will result in the earth's magnetosphere being impacted by the fast moving matter, leading to a geomagnetic storm. This is most likely what occurred when the Earth was recently impacted by a shock wave of plasma.

In conclusion, the Sun seems to go completely dark on April 18th 2018, only 7 days after it went at least partially dark, indicating an accelerating and worsening trend of the Sun weakening, as a result of its interaction with the Planet X System, or System of dead stars that have invaded the Solar System.

References:

[1]   Albers, C. (2018). Article 205: NASA indicates that the Planet X system is affecting the Sun.
[2]   Albers, C. (2017). Article 110: How do stars produce light?
[3]   Albers, C. (2017). Article 26: The Sun goes partially dark on July 5th 2017.
[4]   Albers, C. (2017). Article 60: The Planet X system and the reacting Sun.
[5]   Albers, C. (2018). Article 195: Stellar Cores and the dying Sun.
[6]   Albers, C. (2018). Article 181: Stellar Cores and deciphering gravity.
[7]   Albers, C. (2018). Article 182: Einstein's dream realized: unified field theory of electro gravitation.
[8]   Albers, C. (2018). Article 193: Stellar Cores in the Sun's corona: why do they not collide with the Sun?

Chapter 4

Article 150: Simulated Blood Red Moon

A blood red moon occurred on January 31st 2018. The moon goes blood red when it is in the shadow of the earth. Since red light is less scattered by the earth's atmosphere than the other wavelengths, some red light manages to get through, the atmosphere, and illuminate the earth at the time of the eclipse thus giving the moon a reddish appearance.

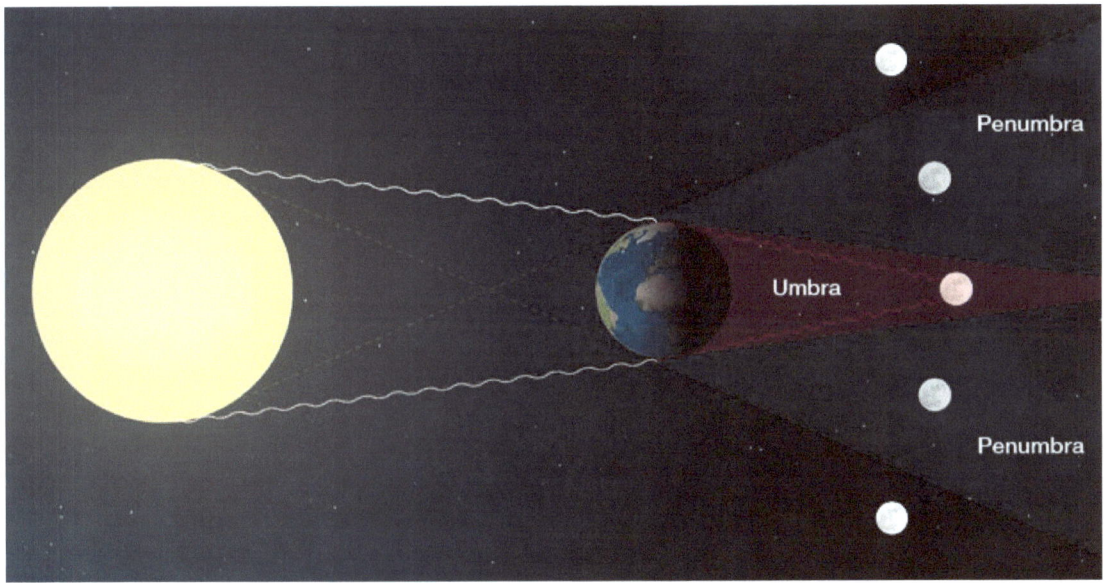

Figure 4.1. The moon turns red during a lunar eclipse when it is in the shadow of the earth. This is because the earth's atmosphere does not scatter red light, as much as other wavelengths, so that some red light coming directly from the Sun manages to get through the atmosphere, and illuminates the moon.

The red sunlight which manages to get through the earth's atmosphere is also refracted or bent by the atmosphere so that it falls onto the moon.

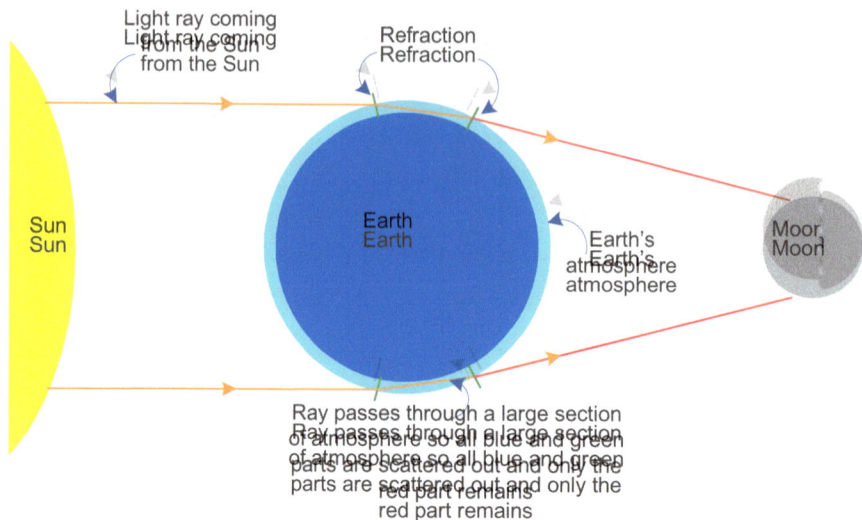

Figure 4.2: The Earth's atmosphere removes or scatters all the light coming from the Sun other than the red light and then refracts that red light so that it falls on the Sun and illuminates it.

However, the footage used by NASA, to show the world what was happening, did not show the real moon. The evidence, that this was not the real moon, lies in the fact, that this moon had a jagged edge, as can be seen in the images below.

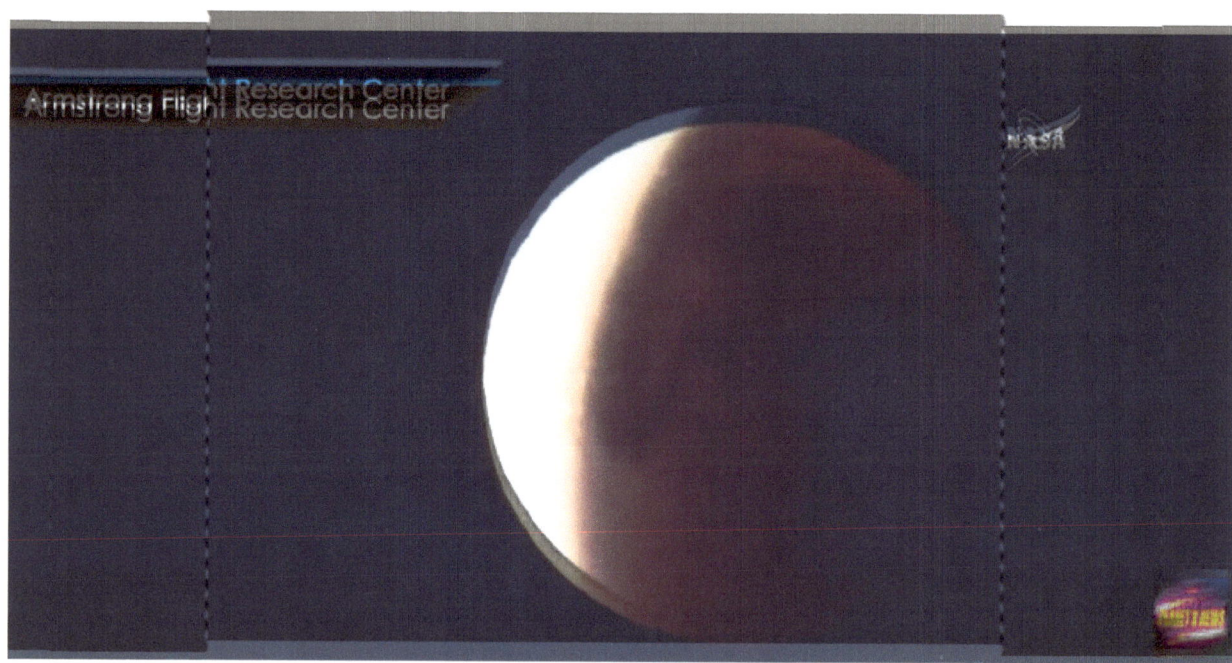

Figure 4.3: A screenshot, of the footage, shown by Planet X News YouTube channel, as provided by NASA, during the lunar eclipse of January 31st 2018. The moon, in the image, has a jagged edge, and cannot thus be the real moon.

Figure 4.4: Jupiter's moon Calisto looks perfectly spherical because the photograph, which was taken, by either Voyager or Galileo spacecraft, from too far away for surface features, to create any significant height change that may produce a non-smooth edge.

The reason, why the moon is not supposed to have a jagged edge is that the real moon is at a distance of 239 000 miles, and at that distance, it is impossible to see any differences in height, at the moon's edge, due to any surface features, on the moon. Thus, the moon is supposed to look perfectly spherical, from earth. This is the exact same reason why Jupiter's moon, Calisto, looks perfectly spherical, in the photograph shown above.

Figure 4.5. The moon, as viewed on January 31st 2018, during the lunar eclipse, clearly has a jagged edge and must therefore be a device, in the earth's atmosphere, which is simulating the moon.

The reason why a simulation device was used, instead of the real moon, is not likely to be that the moon no longer exists, but rather because it is either no longer in the right position, or because the Sun is no longer properly illuminating it, or both. The reason why the Sun may not be properly illuminating the moon is that the Sun has been greatly weakened by the presence of the Planet X System, or Stellar Cores, which have invaded the Solar System. For more details on the fact that the Sun appears to be greatly weakened, see Article 118: Solar activity declining independent of the solar cycle: is the Sun dying? It is likely that both the real Sun, and real moon, are no longer viewed from most place, on the surface of the earth, as detailed in Article 117: Sun simulators below cloud altitude and artificial skies and Article 120: Planet X and the Bible: The artificial Sun simulation system.

In conclusion, a moon simulator was used during the January 31st 2018 lunar eclipse. The reason for the use of a moon simulator is most likely due to an increasingly weak Sun and therefore not able to emit much light and thus not able to illuminate the moon as it once did. The Sun has now been drained by a

System of Stellar Cores, or dead stars, which have invaded the Solar System. It is likely that the real moon, just like the real Sun, is no longer viewed from most positions on the earth's surface.

Chapter 5

Article 151: The Blood Red Moon Simulator at 30 000 feet

In Article 150: Simulated Blood Red Moon I wrote about the fact the moon's jagged edge, seen on the video, provided by NASA, of the lunar eclipse on January 31st 2018, indicated that a moon simulator was being used. However, that article generated some questions, which I would therefore like to address here. First of all, the image I analyzed came from a high definition video, provided by NASA, which Scott shared on his channel, Planet X News, and anyone may go and watch it again. An image of the moon seen in that video appears below. Notice the NASA logo and the words Griffith Observatory:

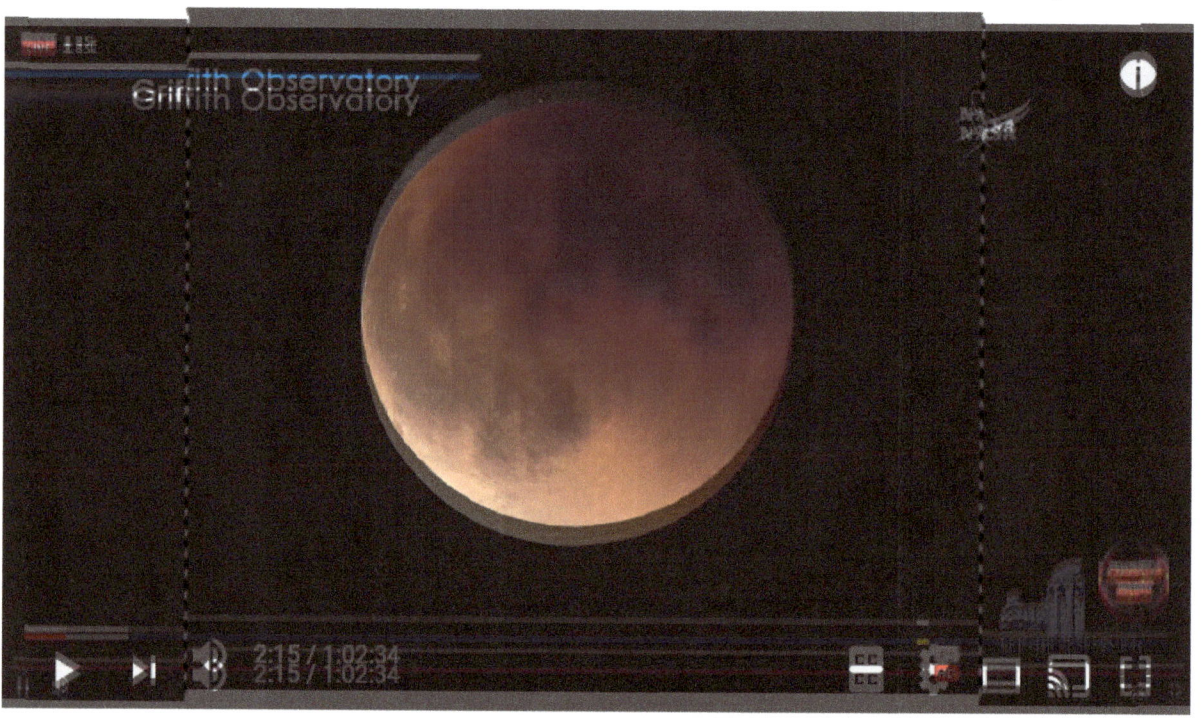

Figure 5.1: Image of the moon taken from the high definition video, provided by NASA.

Now, in figure 2 below, we see an image of the real moon as seen through a telescope. Notice that at least some of its edge is perfectly smooth, and that even places, where it has a jagged appearance the jaggedness is not regular, in any way.

Figure 5.2: Extremely high definition image of the real moon taken through a telescope. It has a very smooth edge, except in some places where a slight difference in height can be seen due to certain moon features appearing close to the edge. This slight difference in height does not produce a regular or repeated step jagged edge.

Some slight differences can be seen at edge on the bottom of the image. The moon does have some very high mountains. The highest of which, if situated, so that it can be seen at the moon's edge, may produce a height change at the edge of 0.6%. However, natural objects like the moon cannot produce repeated height differences along its edge, whatever slight difference there may be, they will all be different from each other and often rounded. A regular jagged edge, exhibiting a repeated pattern, and made with straight lines, is a sign of a manufactured object, not a natural one. It is very difficult to manufacture very large objects, with perfectly smooth curved outlines, so curved outlines are usually made of staggered or stepped straight lines.

Figure 5.3. Close-up of the bottom portion of the moon, from the above real moon image, where features on the moon, close to its edge, produce slight differences in height, at the edge: These do not however have a jagged appearance, in a repeated, or stepped pattern, form.

Figure 5.4. Close up look of the moon, as seen on the high definition NASA video. This moon has a regular jagged edge made out of stepped, straight lines, as you would expect a manufactured object of circular outline, to have. The circular outline is made with staggered, or stepped straight lines. This image is from a point, which is 2 minutes and 15 seconds, into the video.

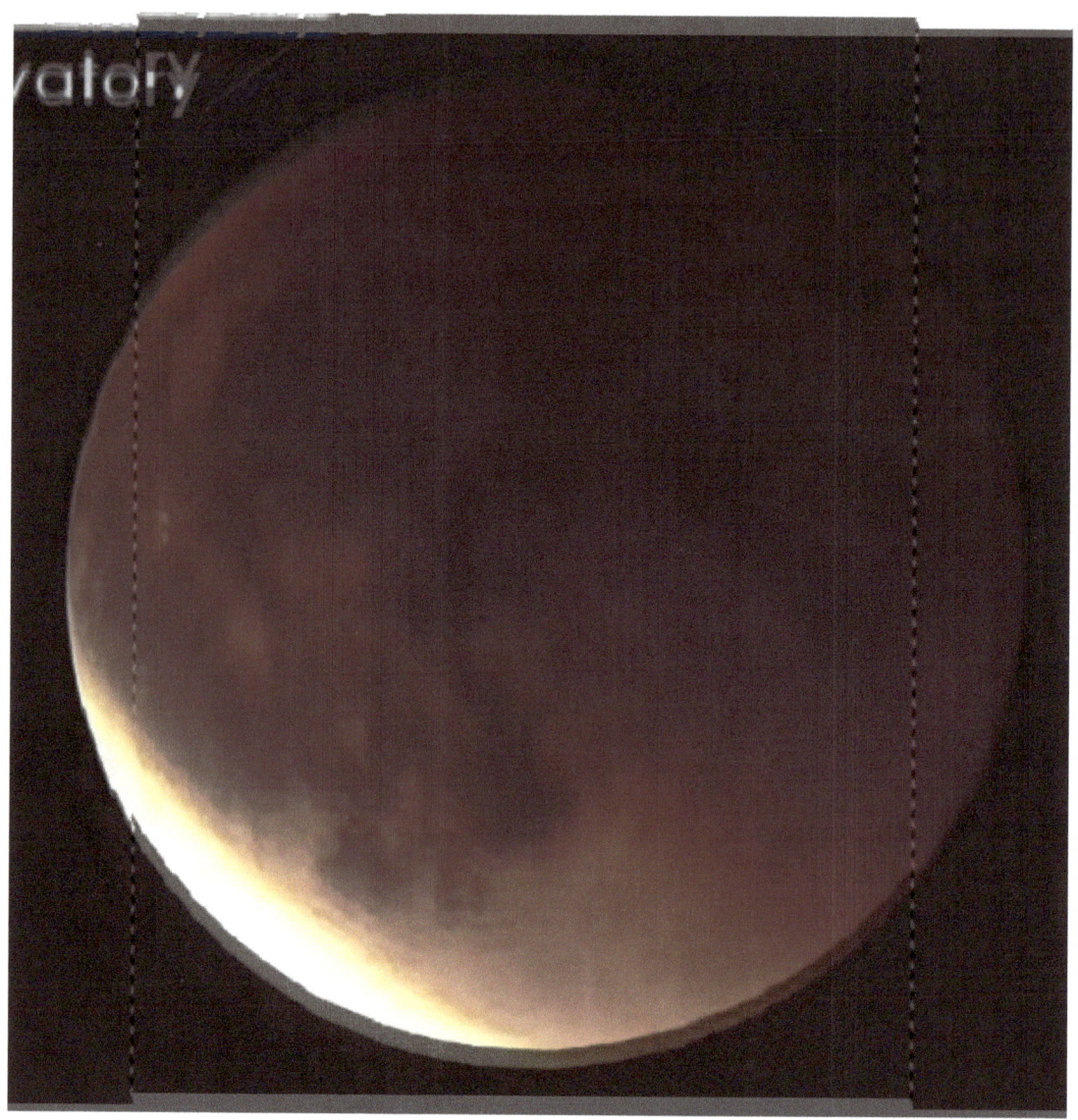

Figure 5.5: Close up look of the moon, as seen on the high definition NASA video. This image is from a point, which is 9 minutes and 27 seconds, into the video. This image has exactly the same jagged edge pattern, made with stepped straight lines, as the image shown in figure 4', from earlier in the video, indicating that this pattern cannot be produced by atmospheric turbulence.

Figure 5.6: The moon's jagged edge is apparent in this close-up, of the left top portion of the image, in figure 4. The jagged edge is stepped and has toothed appearance in places. Only an artificial device can have this type of jagged edge, which is clearly made with straight lines.

Figure 5.7: This jagged edge is stepped, toothed and made with straight lines, indicating that this is an artificially produced device. This is not what a natural object looks like and cannot be the real moon.

Now, the moon has a radius $r_M = 1\,737$ km, and its minimum distance to earth is $R = 363\,104$ km, so the moon's angular width, in the skies, above earth, is

$$\Delta\theta_M = \frac{2r_M}{R}$$

Then, a moon simulator of diameter d, operating at altitude h, will have an angular width of

$$\Delta\theta_S = \frac{d}{h}$$

So, in order for this simulator to appear to have the same size as the real moon, in the sky, it must have the same width. So equating the 2 above equations for angular width, we can obtain an equation for diameter in terms of altitude:

$$\frac{2r_M}{R} = \frac{d}{h} \Rightarrow d = \frac{2r_M}{R}h$$

For a simulator at 30 000 feet, which is equivalent to 10 km, the simulator would thus have to have diameter:

$$d = \frac{2r_M}{R}h = \frac{2(1737)}{363104}\,30000\text{ ft} = 287\text{ ft}$$

The longest Boeing 747 plane is 263 ft in length, so the simulator, if it operates at 30 000 ft, would only have to have a diameter a little longer than that. If it operates at 45 000 ft, its diameter would have to be 430 ft, though.

The photograph, shown in figure 7 below, suggests that the Sun simulator operates, at an altitude lower than the highest altitude, at which clouds form, which is at about 45 000 ft. The same device would most likely operate as a moon simulator, as well. Figure 8 shows a sun simulator device, which is operating at cloud altitude. There are clouds in front and behind it, and its edge is also far from perfectly circular. It is not as easy to discern but it edge is jagged and made out of straight lines as well.

The reason, why these objects do not have a perfectly circular outline is that they are very large and will thus have to be built from lots of different sections, which are then attached to each other, so curved edges are produced with stepped straight lines.

Figure 5.8. Photograph of a late afternoon sun in the sky. The Sun can be seen below a line of dark clouds. Then, there is a break in the clouds. Above the break in the clouds, there is some white cloud and the Sun seems to be shining through these white clouds. However, if the Sun was shining through these clouds, then it should be visible in the break between clouds, and it is not. This means that what seems to be the Sun shining through the clouds is actually a reflection. Thus, the Sun simulator must be below these clouds and its light is being reflected by these clouds. Since the highest altitude clouds form at 45 000 ft, the sun simulator must be below 45 000 ft.

Figure 5.9: The Sun is both behind and in front of clouds, which is only possible, if the Sun is actually an artificial device, simulating the Sun, and operating at cloud altitude. The device is clearly not perfectly circular, but has a jagged outline

Since the Sun and the moon have similar apparent widths from the earth's surface, it should be easy to use the same device to simulate both objects. The moon simply requires a lower intensity of light and parts of the device to sometimes be in complete darkness. The device is most likely a large disk shaped device with perhaps a power source and engines, behind the light emitting side, as illustrated below:

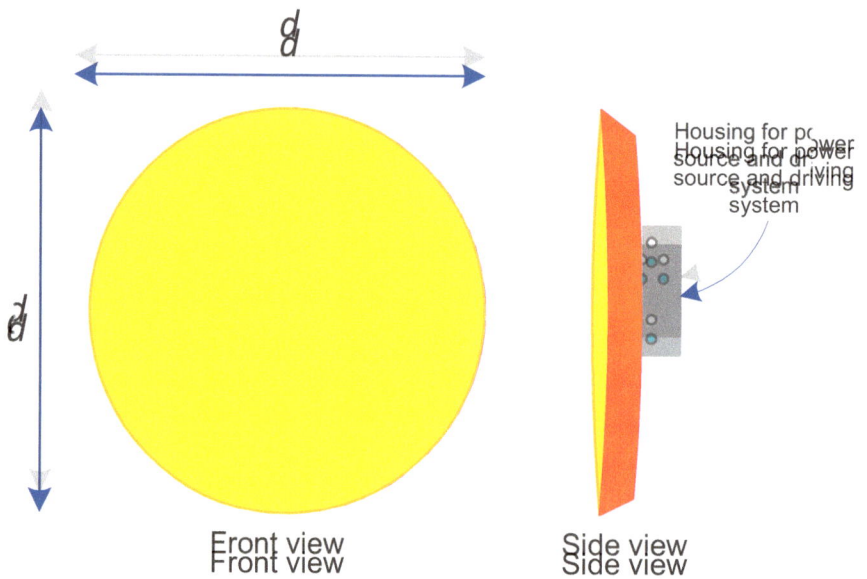

Figure 5.10: Illustration of what a sun/moon simulation device may look like. The diameter, d, will vary between 300 ft, for a device designed to operate at 30 000 ft, to 400 ft, for a device designed to operate at 45 000 ft. The device may also have projection capabilities, so that sun and moon features can appear on the light emission surface.

Such devices may be operating as a part of a grid system, which moves with respect to the ground at a constant speed. Please see the article 123: The biggest scientific cover-up: our Sun, for more details.

In conclusion, a moon simulation device was used in the footage provided by NASA, of the red blood moon event, on January 31st 2018. Such a device, would require a diameter of about 300 feet, if it is designed to work at 30 000 feet.

Chapter 6

Article 160: Sun Simulator: Speeds and Orbits

As I have mentioned in Article 156: Sun simulator behind a cloud [1], and Article 157: Sun simulator: the reasons for its use [2], there are most likely several Sun simulator models, operating at several different altitudes. There seems to be one that operates inside the earth's atmosphere, as this is the only possible explanation, for diverging rays, produced by light interrupted by a cloud as shown below:

Figure 6.1: A photograph of the ocean and sky reveals that the light source is not far behind the clouds producing the diverging ray effect. In addition, this light source is white, not yellow, like the real Sun would be. The pink haze in the sky, above the horizon, reveals the presence of extra natural light sources, extra stars, in the Solar System.

This simulator is most likely disk shaped as shown in figure 2 below. And in order to move, with respect to the earth's surface, at the same speed that we perceive the Sun, to move, with respect to the earth's surface, it would have the same tangential speed, as the earth, as it rotates about its rotational axis. The earth has a radius of 6371 km, so if the Sun simulator is at an altitude of, say, 40 km, its distance from the center of the earth, would be 6411 km, and thus it would have to move at speed:

$$v = \frac{2\pi r}{T} = \frac{2\pi (6411 \text{ km})}{24 \text{ h}} = 1678 \text{ km/h} = 1042 \text{ mi/h} \tag{1}$$

Since it would have to go around the circumference of the earth once, in the time it takes for the earth to do one rotation, or in 24 hours.

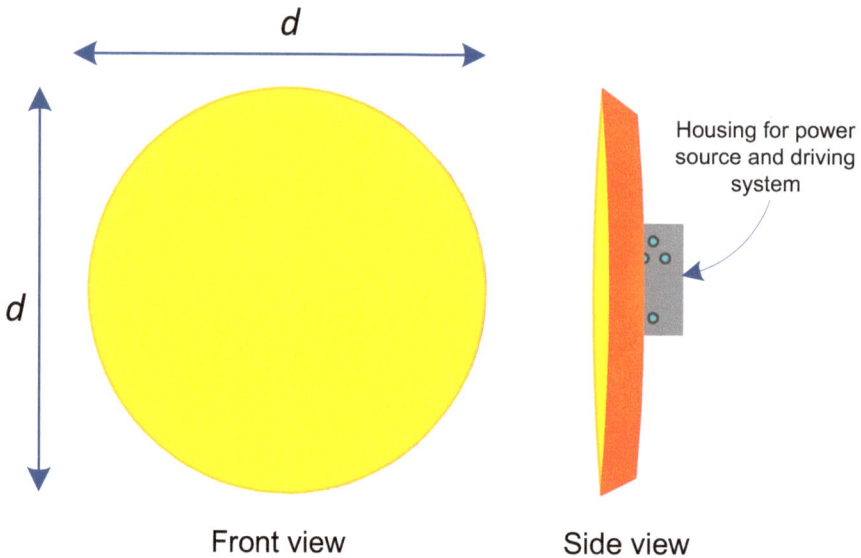

Figure 6.2. Illustration of what a sun/moon simulation device may look like. The diameter, *d*, will vary between 300 ft, for a device designed to operate at 30 000 ft, to 400 ft, for a device designed to operate at 45 000 ft. The device may also have projection capabilities, so that sun and moon features can appear on the light emission surface.

It is likely that the Sun simulators, in the earth's atmosphere, are used in conjunction with Sun simulators, in orbit, and there may be a shift, from one type, to another, in the course of one day, as the one in orbit may provide a better simulation of the Sun, at certain times, such as sunrise, when, for instance, there are no clouds close to the horizon. Now, this Sun simulator would just have to be at the correct orbital altitude in order to move at the correct speed. In order to find that orbital latitude, we just have to realize that it must have an orbital angular speed, which twice as large as the earth's rotational angular speed. We know this because the earth rotates at the rate of one rotation, every 24 hours, and that is what gives the impression of the Sun moving, with respect to an observer on earth, at the rate of one rotation, in 24 hours. Now, a satellite moving with the same angular speed, as the earth, would be a geostationary satellite, and remain over one point above the surface of the earth. So for a satellite to seem to move, with respect to a point, on the surface of the earth, with the same speed that the Sun seems to move, with respect to an observer, on the surface of the earth, the satellite would have to have an angular speed, which is twice the earth's angular speed. Thus, the satellite's angular speed would be twice the earth's angular speed, or

$$\omega_s = 2\omega \tag{2}$$

where ω is the earth's angular speed. Now, we know that a satellite, with an orbital speed equal to the earth's angular speed, would be in a geostationary orbit, which is at an orbital radius of $r = 42\ 164$ km. Thus, resorting to Kepler's third law, we have:

$$\omega^2 = \frac{GM}{r^3} \tag{3}$$

Then, writing equation (3) for the Sun simulator satellite, we would obtain:

$$\omega_s^2 = 4\omega^2 = \frac{GM}{r_s^3} \tag{4}$$

Substituting equation (3) into (4) we get:

$$4\frac{GM}{r^3} = \frac{GM}{r_s^3} \Rightarrow r_s = \frac{r}{\sqrt[3]{4}} = 26\ 568\ \text{km} = 16\ 500\ \text{mi} \tag{5}$$

If we subtract, the earth's radius, from the Sun Simulator satellite orbital radius, we obtain the following orbital altitude:

$$h_s = r_s - R = 20\ 197\ \text{km} = 12\ 543\ \text{mi}$$

Now, this type of Sun simulator would not need to have a means to propel itself, like the one in the atmosphere, but on the other hand, it would have to be extremely large, and powerful, in order to be able to produce enough light, and heat, to simulate the Sun, from such a high altitude. It is thus probably much easier to simulate the Sun from within the atmosphere.

Figure 6.3. ISS Sun simulator in operation: It has what seems to be laser beams focused on it from above and below.

There may be another type of Sun simulator, which may operate, at a lower orbit, such as the ISS orbital altitude of 400 km, or 240 mi. This Sun simulator would however be moving too fast to properly simulate the Sun, for an observer on earth, but it may be used in ISS footage, and it may have holographic projection capabilities, which can circumvent the high speed problem. A design of a Sun simulator, based on photographs of this simulator, is detailed in Article 58: The ISS Sun Simulator [3], and one of the design diagrams appears in figure 4 below.

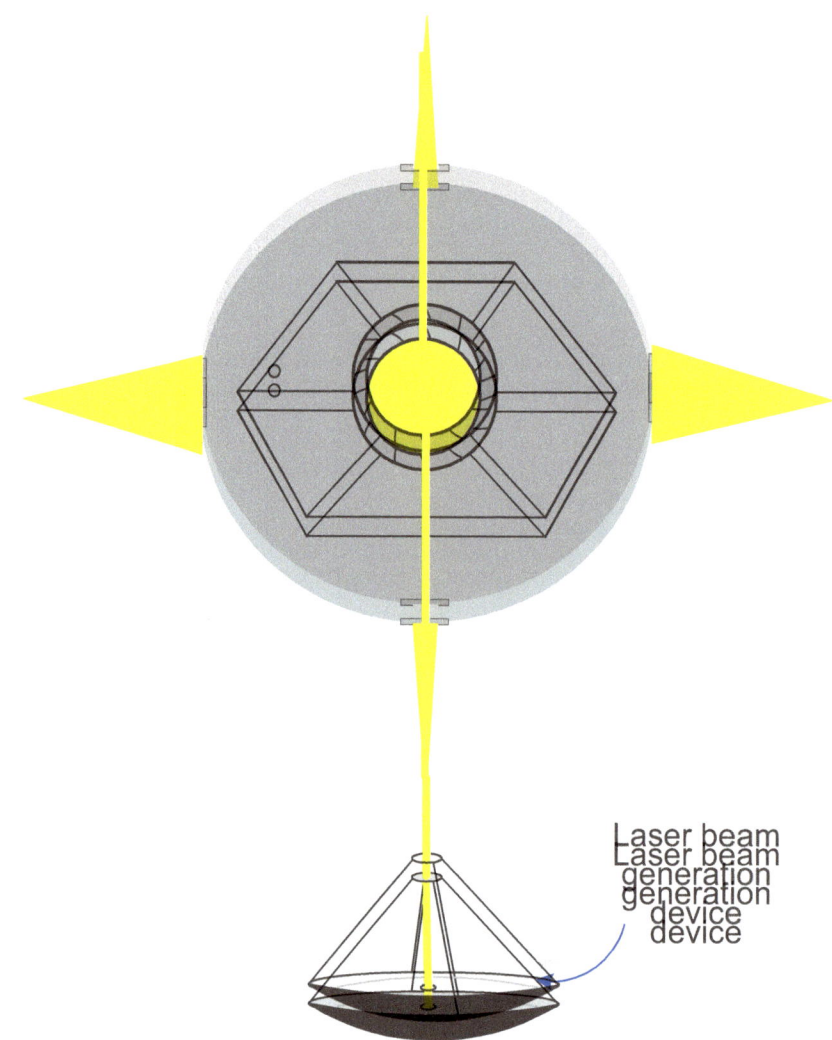

Figure 6.4: The ISS Sun Simulator, with a laser beam generation device, below it, giving rise to the vertical light beam, the device is observed to have. Another similar device would be positioned above the device in order to produce the top vertical beam.

In conclusion, there are most likely several Sun Simulator designs in operation. The one in the earth's atmosphere would have to move with a speed of just over 1000 mi/h (1700 km/h) and a Sun Simulator in orbit would have the correct speed at an orbital altitude of 12 500 mi (20 200 km)

References:

[1]   Albers, E. (2018): Articles 156: Sun simulator behind a cloud.
[2]   Albers, E. (2018): Article 157: Sun simulator: the reasons for its use.
[3]   Albers, E. (2017): Article 58: The ISS Sun Simulator.

Chapter 7
Article 164: Secret advanced technology is being used in our skies

I came across a video posted by the YouTube channel: William Mignoli [1] where instead of an airplane with a white aerosol trail behind it, we see a craft, which seems to be immersed in a light emitting field. Two screenshots from this video appear below.

Figure 7.1: Instead of the normal chemtrail airplane leaving a white trail behind it, we see a craft surrounded by a field. The most likely purpose of the field is to cloak the craft and make it appear to be a normal airplane. The cloaking device must have malfunctioned in this instance.

The field seems to be filamentary (producing string like structures) in nature, which is indicative of a high energy electric field. The most likely purpose would be to cloak the craft. Since chemtrail craft, seen in our skies, appear to be in the shape of airplanes, this strange capture seems to suggest that this craft would also appear to be a plane, if the cloaking device had not malfunctioned.

Figure 7.2. Screenshot from a video showing a strangely shaped craft spreading aerosols in earth's atmosphere. The shape seems to be as a result of a strong magnetic field around it, which changes in the video footage, and at times seems to split into two, with the back part larger than the front part, as seen here.

This means that more or possibly even most chemtrail airplanes we, see in our skies, are not actually normal airplanes but craft cloaked and made to look like airplanes. They may even be a type of drone, which removes the human element who may develop a conscious regarding the spraying of fellow humans as if they are an insect plague to get to rid of. It also indicates that advanced technology, kept secret from the general population, is being used in our skies.

The chemtrail or aerosol injection program seems to have two main purposes: 1. to hide the presence of Stellar Core System, which has invaded the Solar System and come to the Sun's Corona to absorb energy from the Sun (see Article 116: Planet X Objects: unbelievable evidence and size [2]), from the earth's population. 2. To poison every living being on the planet as detailed in Article 148: The purpose and effect of chemtrails [3].

In conclusion, advanced technology, such as craft with cloaking capabilities, which is also being kept secret from the earth's population, is being used in the earth's atmosphere.

References:

[1]   Migndi, W. (2018). Electron Magnetic Insect Machine Chemtrail Plane Object. https://www.youtube.com/watch?v=ayzceFgHYzA&feature=youtu.be
[2]   Albers, C. (2017). Article 116: Planet X Objects: unbelievable evidence and size.
[3]   Albers, C. (2018). Article 148: The purpose and effect of chemtrails.

Chapter 8

Article 165: Sun Simulator: irrefutable evidence

In Article 156: Sun Simulator behind a cloud [1], I showed that Sun simulators are being used in the earth's atmosphere. The argument, I used to show this, was based on the fact that diverging rays can only come from a source, which is close to the earth's surface, and cannot be the real Sun. But it seems that, judging by some of the comments left under my videos dealing with this, that some people, even some who claim to know how to use a telescope, have not been able to understand that argument. So in this article, I am going to use another photograph, and I am going to make it even clearer. The photograph I will use is shown in figure 1 below. Now, the first course I ever taught, at university, was on Geometric Optics, and in order to teach this course properly, I had to understand the principles involved really well, as well as find ways to clearly teach the concepts, so to me this is a good challenge.

Figure 8.1. Light from the light source in the sky pokes through the clouds in different directions indicating a light source just above the height of the clouds.

Now, the most important concept in Geometric Optics, which most of us intuitively know is that **light travels in straight line**s. This is how we are able to locate light sources in our environment. If we have a lamp in our room, we see its brightness and are able to reach out to it. This is also how we are able to locate the Sun in the sky. If we look up at the sky, during the day time, and see a very bright light in the sky, we are then able to deduce where the Sun is, in relation to our position on the surface of the earth, as illustrated in figure 2 below. Because we are very far away from the Sun, its position in the sky does not change, even if we move several miles on the surface of the earth. However, if we move in relation, to a lamp in our room, the lamp will seem to change positions because we are so close to it. The fact that the Sun does not move when we move tells us that it is extremely far away, and the fact that lamp seems to move if we just move a small distance tells us that it is very close to us.

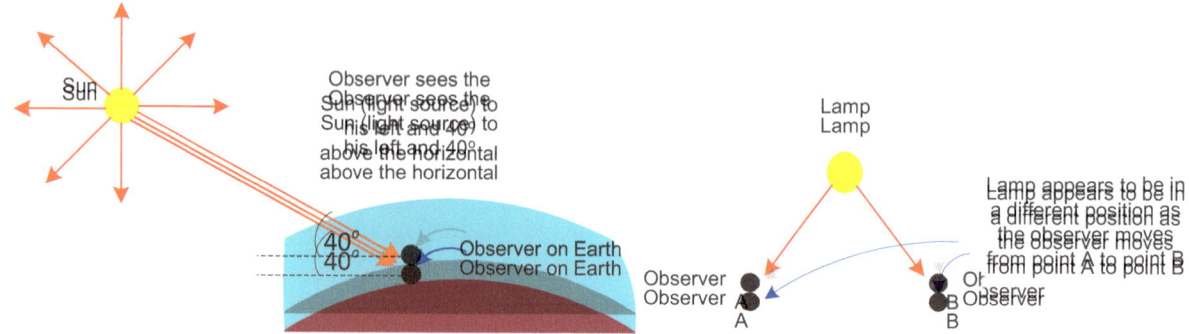

Figure 8.2: We are able to locate light sources because light travels in straight lines. We can also determine how far the light source is according to how much it seems to move when we move some distance in relation to it. Left: The Sun is extremely far away, so its position does not change, in relation to the observer, even if the observer moves several miles. Right: In the case of a lamp, in the observer's room, for example, the lamp appears to be in different positions if the observer moves only a small distance.

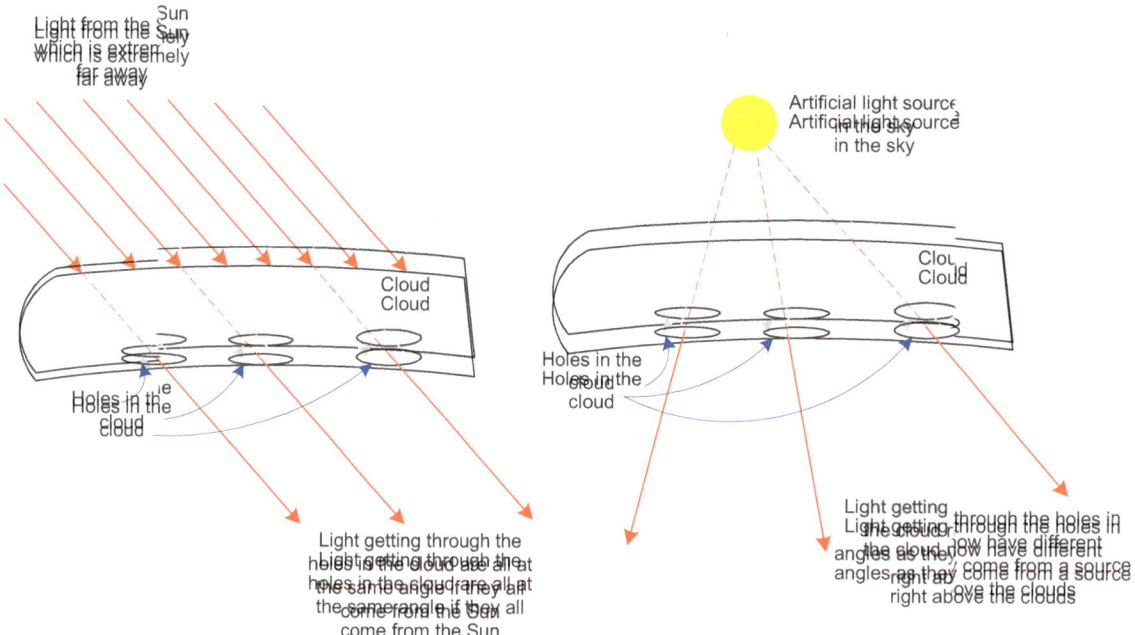

Figure 8.3: If the light source is the Sun, light coming through holes in the cloud will be parallel or at the same angle. But if the light source is just above the clouds then the light rays, coming through the holes in the cloud, will be at different angles.

Figure 3 illustrates the difference between light coming from the Sun, which is a light source at infinity, and a source of light in the earth's atmosphere. The rays of light coming through holes in the cloud can only be at different angles, if the light source is just above the clouds. If the light source is the Sun, the light rays will all come through the holes, in the clouds, **at the same angle**. We can also find the source of the light, by extending the light rays coming through the different holes in the clouds, backwards:

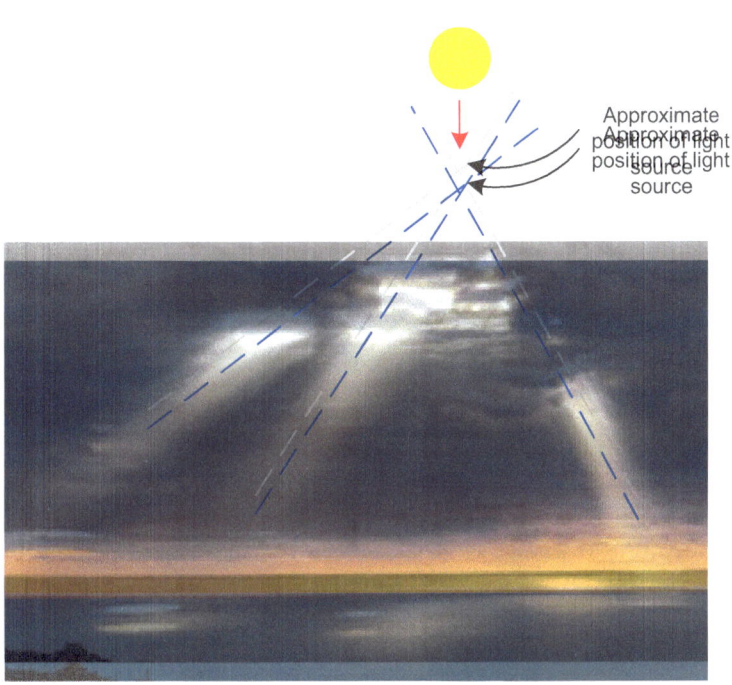

Figure 8.4: Extending the rays of light coming through holes in the cloud in this photograph allows the position of the light source to be determined. This position is not far above the clouds. This means that the source is artificial, i.e. a Sun Simulator. This light source is sufficiently far away not to move if the observer moves a short distance and it is thus only when its light is interrupted by a cloud that its proximity and position in the atmosphere becomes obvious.

Figure 5 illustrates how the light source's position can be found by simply extending the light rays coming through holes in the cloud backwards. The position of the light source is not far above the clouds and cannot therefore be the real Sun, which is 93 million miles away. This is irrefutable proof that an artificial device is being used to simulate the Sun.

Notice, in addition, that the light in the sky, above the horizon, is dark yellow and orange. This light cannot be coming from the light source in the sky as the light it emits is quite different, nor would it be able to illuminate the earth's atmosphere from its position within the atmosphere. In addition, using the fact that the Sun simulator will follow the Sun's position, and since this light source is way above the horizon, it is not possible that the real Sun's light would be so dark yellow or orange. This indicates that a light source outside the earth's atmosphere, a natural light source, and therefore a star is emitting dark yellow or orange light and illuminating the earth's atmosphere with this light. The presence of these natural light sources, stars, in the sky, as well as their weakening effect on the real Sun, presents therefore a reason for the use of artificial devices to simulate the Sun, by those who desire to keep the truth, from the earth's population.

In conclusion, the analysis of a photograph, showing the Sun poking through holes in the cloud, based on the simple principle that **light travels in straight lines**, a principle that we all use since early childhood to locate light sources, leads to the conclusion that the light source is in the earth's atmosphere and cannot therefore be the real Sun. This constitutes irrefutable evidence that an artificial device, a Sun simulator, is being used to simulate the real Sun.

Chapter 9

Article 123: The biggest scientific cover up: our Sun

The evidence that the Sun can and does go dark, periodically, started, in my view, with SDO images, which reveal that the Sun goes dark during the so called SDO eclipse season. I describe this evidence in article 110: How do stars produce light? [1]. The SDO eclipse season has yielded SDO images showing the Sun going dark since 2011. However, the Sun does not just go completely dark, it is being drained by the presence of the Planet X Objects or Stellar Cores, in the Sun's corona. These Stellar Cores have been arriving at the Sun's corona, for at least 100 years. I have been writing about these objects for many months now, one of the most recent articles discussing the visual evidence for the presence of these objects in the Sun's corona is article 116: Planet X Objects: unbelievable evidence and size [2]. However, it has now come to light that the Sun may be emitting much less light, than I even thought, and that this is being covered up. The left image in figure 1 below shows a Sun, which is almost completely dark, and on the right is an image where it does not look dark. This indicates that the image was modified in order to give the perception that the Sun is not as dark as it really is in this wavelength.

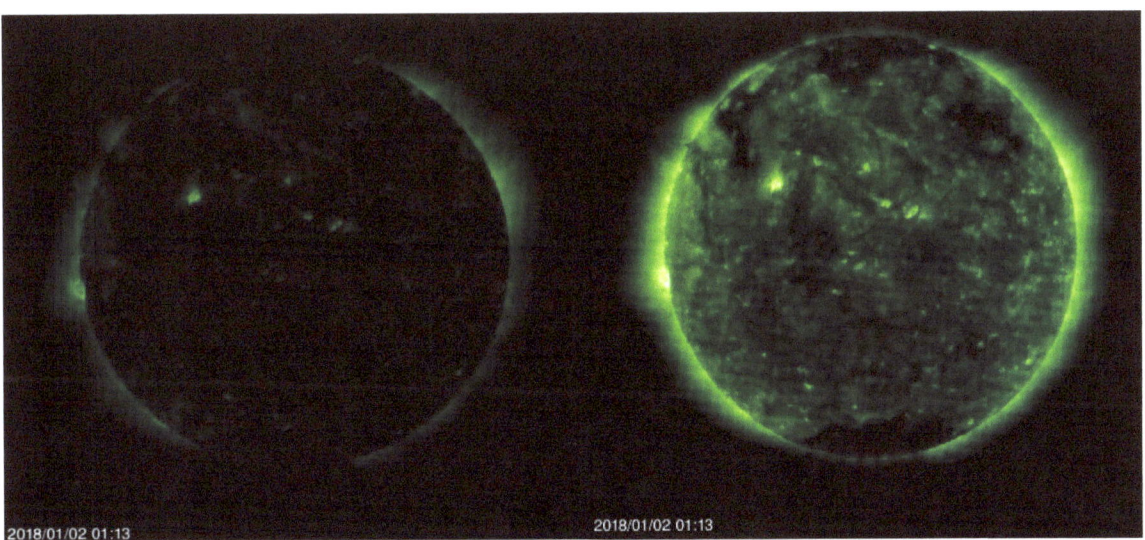

Figure 9.1. Stereo A images of the Sun in ultraviolet light from January 2nd 2018 at 1:13 (UTC). The one on the left was on the ISWA (Integrated Space weather) website and the one on the right, was taken directly from the Stereo website. In the one on the left it is obvious that the intensity of the light emitted by the Sun in this wavelength is extremely low. In the right image the Sun looks normal indicating that much brighter color was assigned to very low intensities.

In ultraviolet images, certain colors are assigned to a particular wavelength, and different tones of that color are assigned to different intensities, with the lightest or brightest tones being assigned to the highest intensities. Intensity is a measure of the number of photons, of a certain wavelength, emitted. Thus, the higher the intensity of the light emitted by the Sun, the more photons, with a particular intensity, are emitted by the Sun. The two images shown in figure 1 show that bright green tones usually assigned to normal high intensities are being reassigned to very low intensities, so that instead of looking very dark, due to a dramatic drop in light intensity emission, the sun looks as if it is emitting light at a normal intensity, thus giving a false perception of the Sun's state. If this is being done in this wavelength, it is most likely being done to all wavelengths detected by the various satellites:

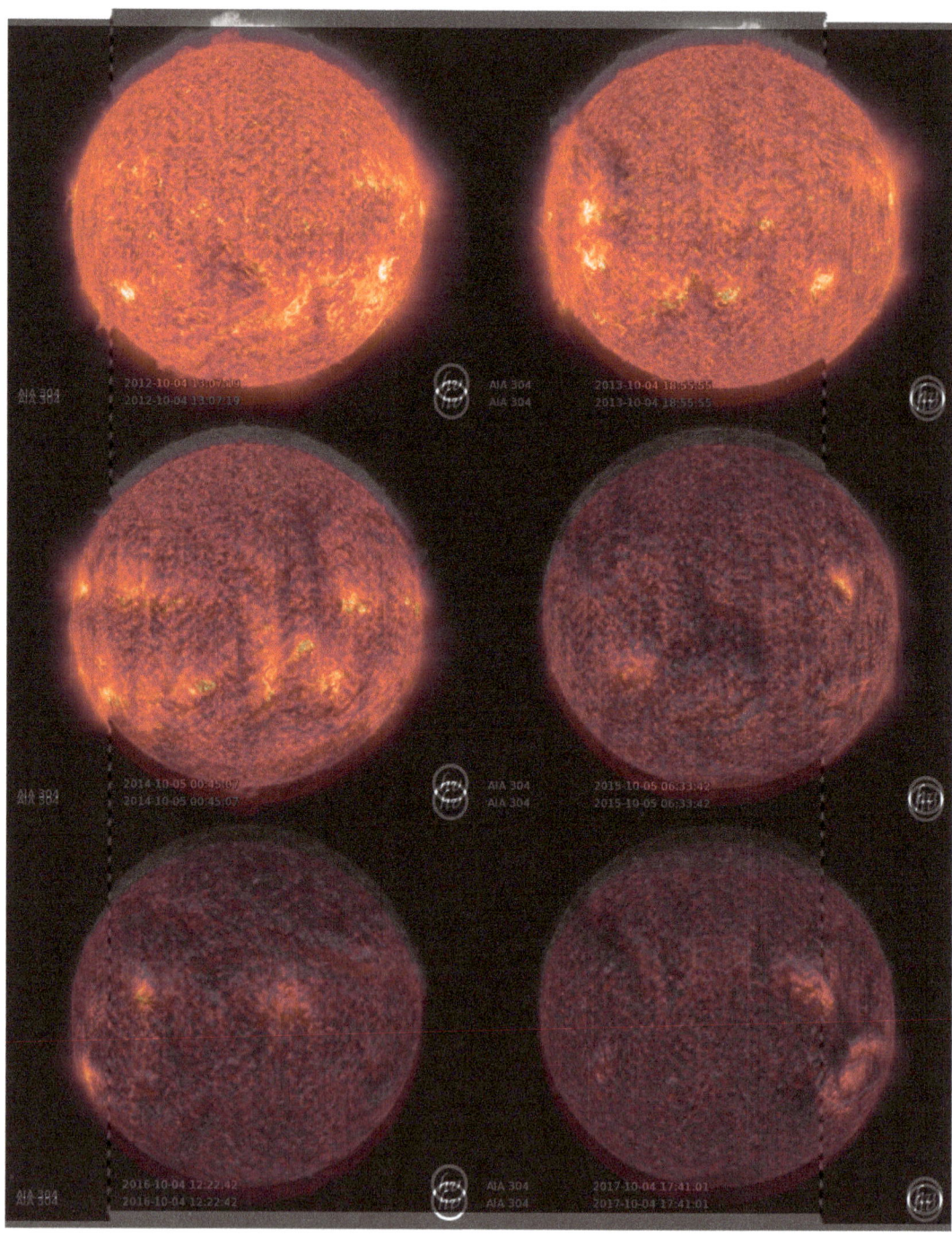

Figure 9.2: SDO images of the Sun in the 304 angstrom wavelength from 2012, 2013, 2014, 2015, 2016 and 2017. The Sun has clearly grown darker in this wavelength over the years. Comparison of the brightest spots indicates in the different images; show that they can achieve the same lighter color, and

that therefore the increased darkness is not due to a change in the color assigned to different intensities. The 304 angstrom wavelength is emitted mostly from the upper chromosphere and this is therefore the Sun's layer that is seen in these images. The weakening effect seems to have accelerated in the last 2 years.

Figure 2 shows that the Sun has weakened significantly since 2012 and independent of the solar cycle. The fact that the percentage of the Sun covered in coronal holes is increasing, that solar maximums are late and that signs of solar minimums are arriving early are additional signs that the Sun has been weakening. In addition the Sun's magnetic field which is stronger in sunspots than anywhere else on the Sun's surface was shown to decline between 1996 and 2009 by about 50 gauss per year and thus independently of the solar cycle. This is discussed in the article 118: Solar activity declining independent of the solar cycle: is the Sun dying? [3].

In the light of what figure 1 suggests is being done to cover up the Sun's true state, it is likely that the weakening indicated in figure 2 is much worse and it is possible that this light emission is in fact not just very weak but going up and down over hours and days.

Figure 9.3: A photograph of one of the sun simulation devices in operation in the earth's atmosphere. It has a hexagonal outline and seems to be made up of an array of hexagonal mirrors and may be of a design similar to what that shown in figure 4.

The fact that brighter color was reassigned to at least one Stereo image of the Sun, in order to make the Sun appear to be emitting at a much higher intensity than it actually was, and that there are vast amounts of data deleted, in terms of hours of images, that are simply not available or deleted, suggests that the Sun may be even weaker than the data that is available suggests. This is likely one of the reasons why we no longer seem to see the natural Sun anymore, but instead, we see sunlight produced by an artificial device, as the one seen in the photograph shown above. It is also extremely likely that not only the Sun, but also the moon and planets, are being simulated, by a system of devices, which move across the surface of the earth, part of which is operating in the earth's atmosphere, and another part is operating from orbital satellites. I discussed the evidence for the use of a Sun simulation system in article 121: Planet X and the Bible: The artificial Sun simulation system [4].The type of system that may be operating may be made up of hundreds of sun simulators, possibly a system of 6 equally spaced systems, assigned to a certain latitude. The system would move at constant speed with respect to the earth's surface and different parts would be switched on and off depending on whether day or night or different parts of the day need to be simulated. This system is illustrated in figures 4 and 5 below.

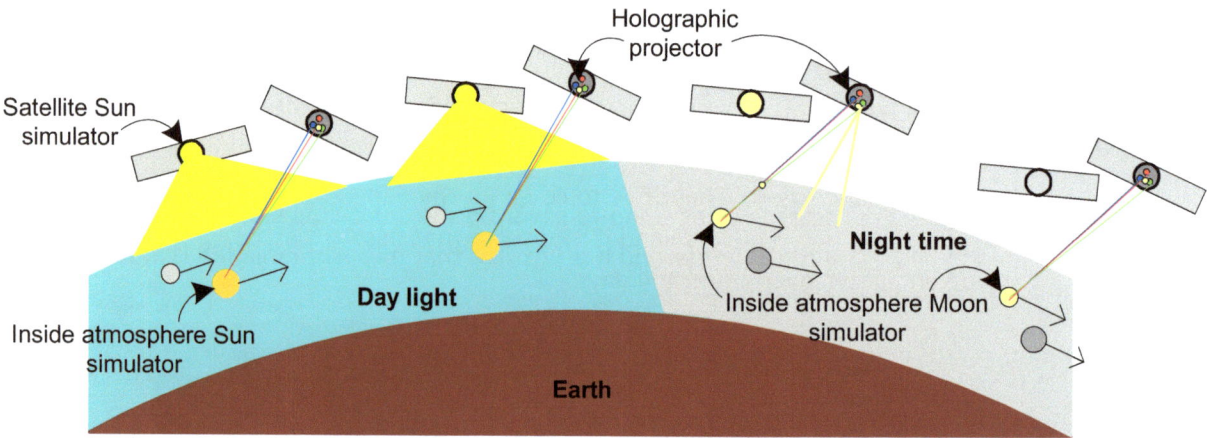

Figure 9.4. Illustration of the artificial Sun simulation system made up of elements in the earth's atmosphere and other outside it. The system moves slowly and parallel to the earth's surface so if a part of it does not work it will be noticed when it moves across a section of the earth's surface. The sun simulation elements are turned on during the day and the moon and planet elements are turned on during the night or whenever needed.

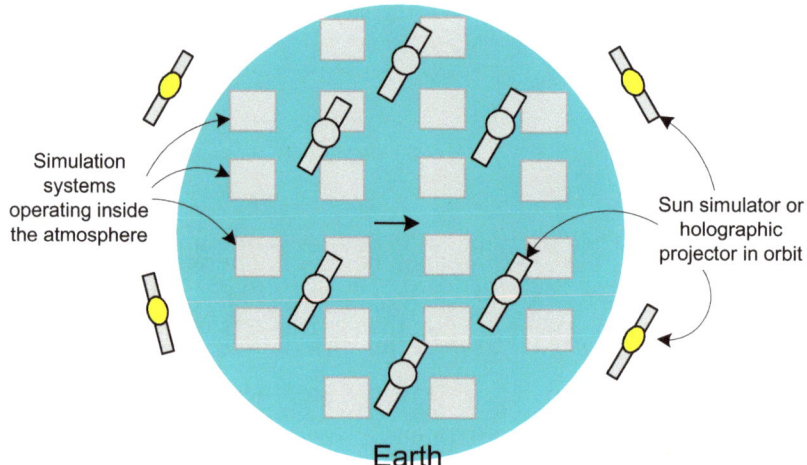

Figure 9.5. From space, the Sun simulation system is made up of many subsystems which slowly move across the surface of the earth in a grid pattern. Several components will, most likely, be in orbit and thus operate from satellites. Each grey square may be made up of a sun simulator, a moon simulator, a holographic projector and a lens system. The arrow indicates the direction of motion.

In conclusion, it appears that the Sun may be much weaker and darker than the images we are getting from solar observing satellites and spacecraft seem to indicate and that the Sun's true state is being covered up, whilst a Sun simulation system, is being used to hide this fact from the earth's population. This is indeed the greatest scientific cover up in the history of mankind.

References:

[1] Albers, C. (2017). Article 110: How do stars produce light?
[2] Albers, C. (2017). Article 116: Planet X Objects: unbelievable evidence and size.
[3] Albers, C. (2017). Article 118: Solar activity declining independent of the solar cycle: is the Sun dying?
[4] Albers, C. (2017). Article 121: Planet X and the Bible: The artificial Sun simulation system.

Chapter 10

Article 166: Sun Simulator and lens system

In Article 165: Sun Simulator: irrefutable evidence [1], I clearly showed, based on the concept that light travels in straight lines, that a Sun Simulation device is being used in the earth's atmosphere. This can be seen from the fact that light rays, going through 3 different holes, in a thick cloud emerge at very different angles. These diverging beams can only be produced by a light source, which cannot be far above the cloud, and thus has to be in the earth's atmosphere.

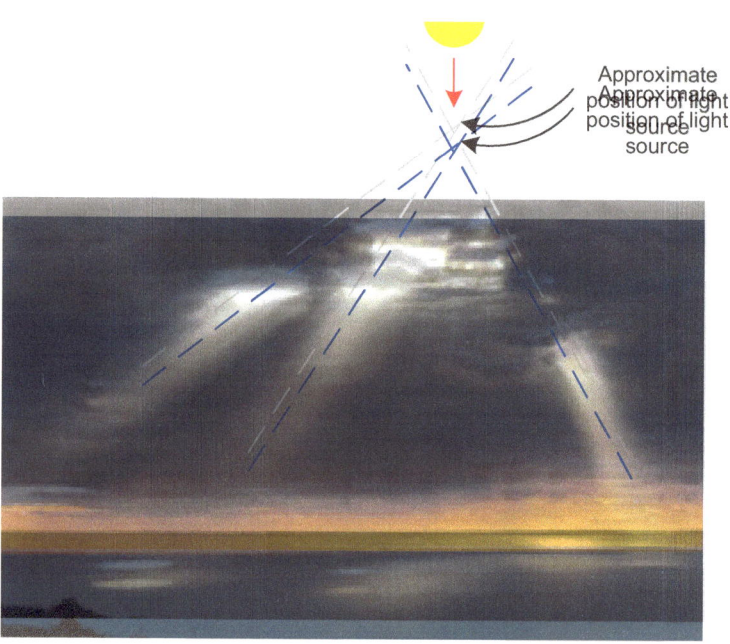

Figure 10.1: Diverging light rays produced by large holes in a thick cloud indicate that the light source is not at infinity but close. Extending the rays of light, coming through holes in the cloud, in this photograph, allows the position of the light source to be determined. This position is not far above the clouds. This means that the source is artificial, i.e. a Sun Simulator. This light source is sufficiently far away not to move, if the observer moves a short distance, and it is thus only when its light is interrupted, by a cloud, that its proximity, and position, in the atmosphere, becomes obvious.

In this article, I will show that these Sun Simulation devices are most likely being used with lens systems in the earth's atmosphere. The need for a lens system is illustrated in figure 2 below. Let us suppose, for the sake of simplicity, that the Sun is in its midday position and a simulator is used in the Earth's atmosphere to occult, or appear in front of, the Sun. Now, unless the observer is exactly beneath the simulator, the light from the Sun and the simulator will not appear to come, from the same point, in the sky, unless a lens system is placed below the simulator, in a circular arrangement. This will however not be necessary, or possibly not even work, if there is thick cloud cover, which is why the closeness of the simulator may then become apparent, as in the photograph shown in figure 1, when holes in the cloud appear.

Figures 2 and 3 below illustrate the problem that occurs when no lens system is used. The Sun is not properly occulted, unless the observer is in the exact position beneath the simulator. Due to the Simulator's proximity to the surface of the Earth, it also becomes necessary to use a whole system of simulators operating, each operating over its own region.

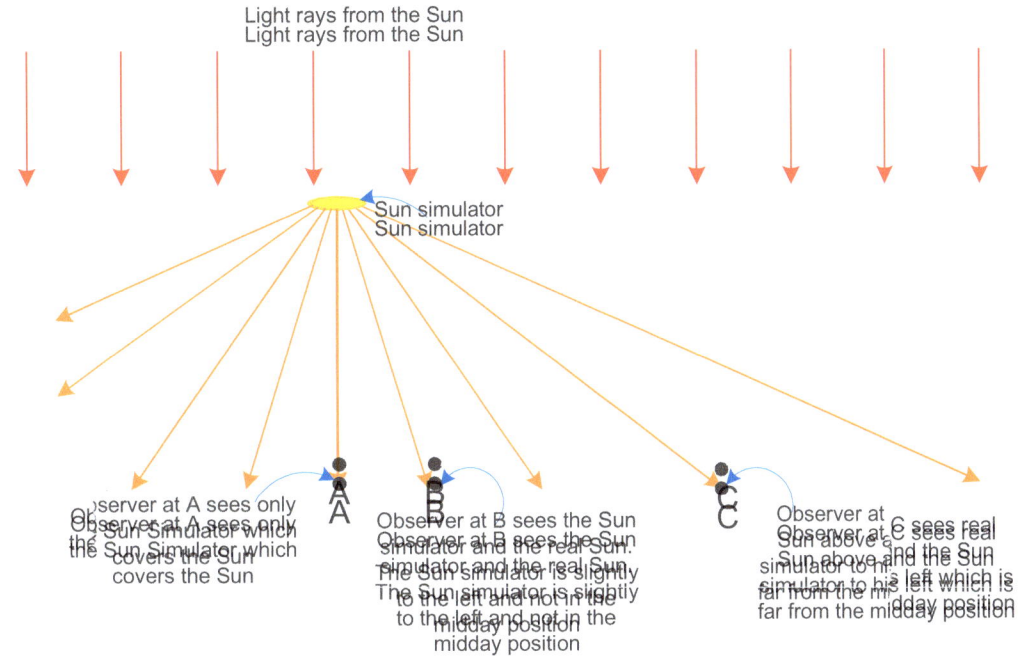

Figure 10.2: The Sun Simulator is in the correct position to occult the sun for observer at A, but not for observers at B or C. At times when the Sun is dark, or eclipsed, only the Sun Simulator would be visible but it would appear to be in the wrong position. This would most likely not be noticed by the observer at point B but it becomes more noticeable for the observer at C.

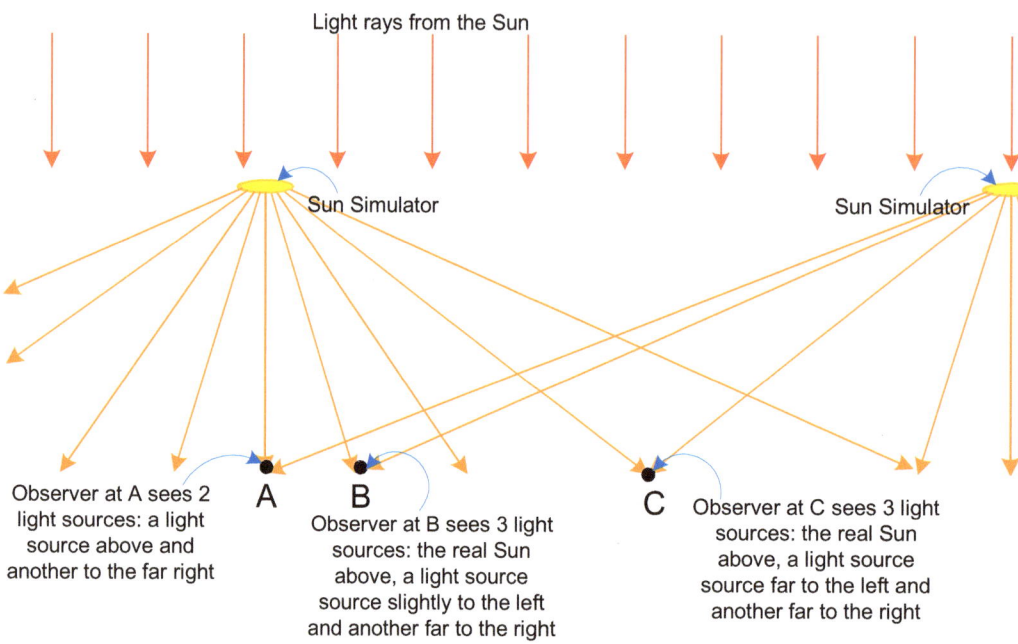

Figure 10.3. As a single Sun Simulator would eventually seem to be too low below the horizon or might not even be visible anymore due to the earth's curvature, more than one Simulator may be required. This would then lead to the possibility that all observers see more than one light source.

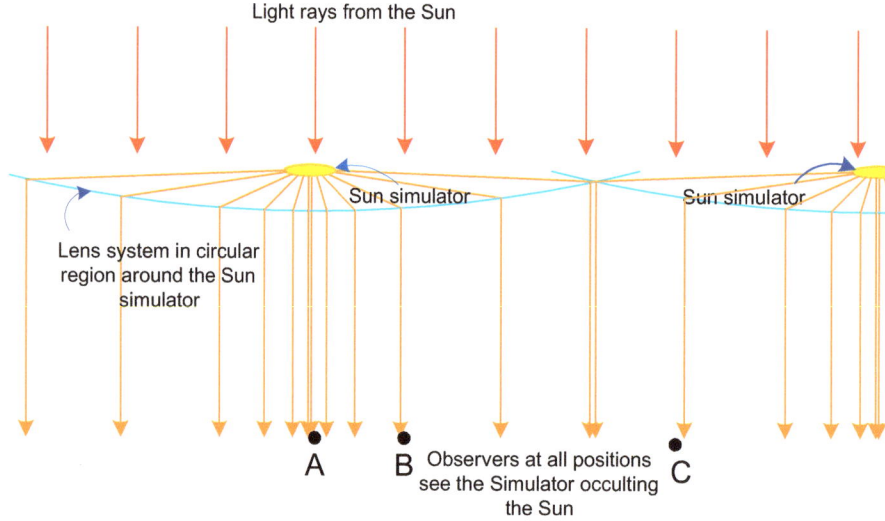

Figure 10.4. With a lens system around each simulator all observers see the Simulator occulting the Sun.

Figure 4 shows how a lens system positioned below, and centered on, each Sun Simulator will result in the real Sun being perfectly occulted for all observers. If the real Sun is lower in the sky. The Sun lens system can be used to change the angle of refraction so that it emerges with the correct orientation.

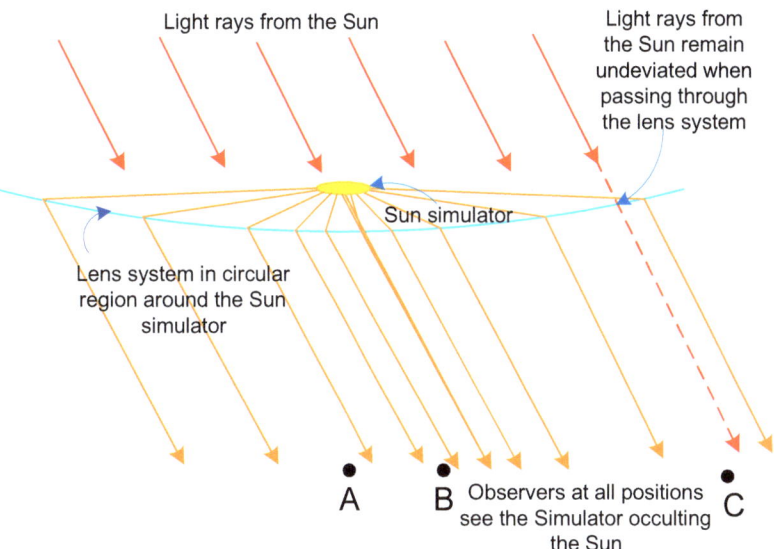

Figure 10.5. The lens system deviates light rays coming from the Sun simulator to the correct angle so that all observers in its region of operation still see the Sun occulted by the artificial light source. The lens system will however need to incorporate the reflection capabilities in the case of large angles of deviation.

It will not be necessary to cover the whole surface of the earth with the lens systems however. They are not necessary over regions with thick cloud cover, over the ocean, except over known shipping lanes and airplane routes and unpopulated areas such as large desert areas.

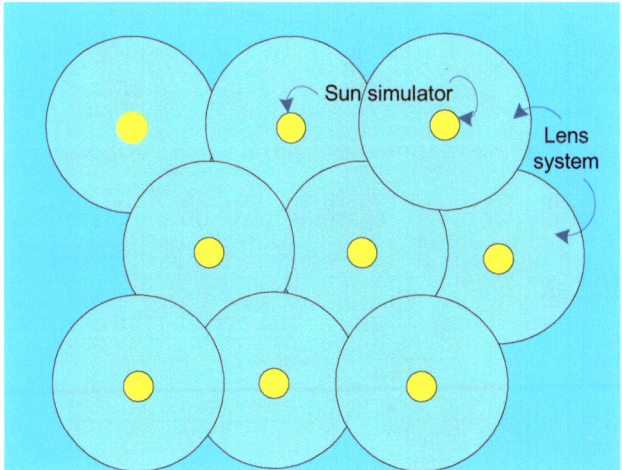

Figure 10.6. Illustration of the likely view, from space, of the Sun simulator and lens system, over a highly populated region of the Earth's surface.

If one large spherical simulator is in operation in the correct orbital position, or at an orbital altitude of 12 500 miles, as detailed in Article 160: Sun Simulator: Speeds and Orbits [2], so it orbits the earth at the same speed that the Sun seems to move across earth's skies, only regions corresponding to close to Sunset and Sunrise position, as well as regions close to the North and South Poles will require additional

Simulators in the atmosphere. This is illustrated below. This Sun Simulator would have to have a diameter of 200 km in order to appear to have the same angular width as the Sun in the sky, from the earth's surface. However, it is possible that another system of lenses in operation around it, i.e. from the same orbital altitude may be able to magnify it, so that it does not have to be that large.

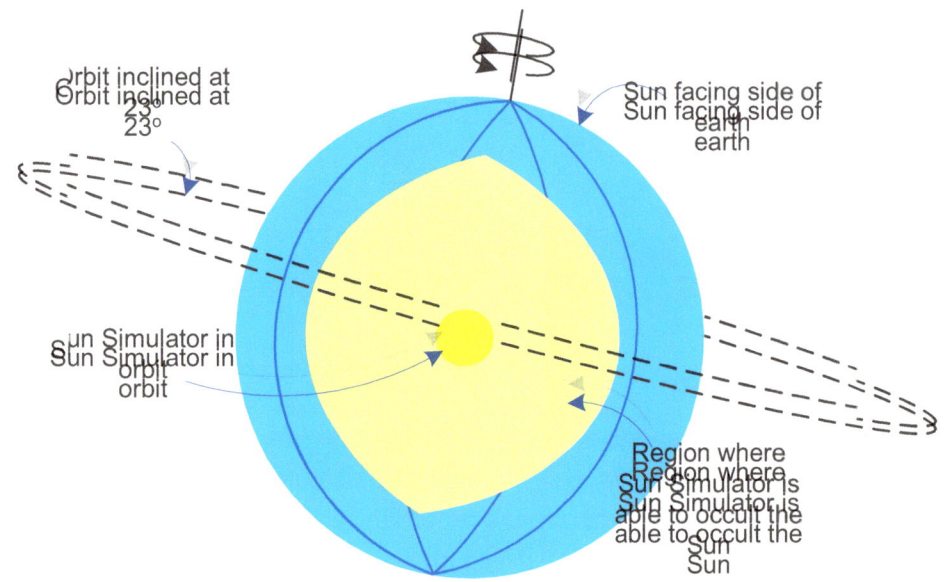

Figure 10.7: The blue region represents the Sun facing side of planet Earth. The Sun Simulator maintains an orbital altitude (12 500 mi) and corresponding speed, such that it is always over a position, on the surface of the Earth, above the Equator, which corresponds to midday.

This Sun simulator, in orbit, will have to be extremely large, but it will be able to, in conjunction with lens systems, where necessary, to occult the Sun over a region on the surface of the earth centered on the midday position at the equator. Areas on the Sun facing side of Earth, East, West, North and South of this region will most likely require Sun Simulators in the Earth's atmosphere.

Another question, which usually crops up regarding the Sun Simulator, revolves around the energy source to power it. There is a huge amount of energy in the Earth's ionosphere and at cloud altitude, which also comes from the ionosphere. This power source in unlimited as it is continuously being replenished by the Solar Wind, and it may be tapped to power all the devices in this system. High energy lasers may be used to connect the source of the power, i.e. the ionosphere, and the different devices, including the Sun Simulator in orbit. High energy lasers ionize air so that it is able to conduct electricity and can thus operate like wiring in a circuit.

In conclusion, as shown in previous articles [1] artificial light sources, which simulate the real Sun, are being used in the earth's atmosphere. These Sun Simulators with the possible addition of one large simulator, operating from orbit, need to operate in conjunction with advanced lens systems located in the earth's atmosphere.

References:

[1] Albers, C. (2018). Article 160: Sun Simulator: Speeds and Orbits.
[2] Albers, C. (2018). Article 165: Sun Simulator: irrefutable evidence.

Chapter 11

Article 213: Producing lenses out of air in the earth's atmosphere

In a recent video, Jeff P, showed video footage, sent to him, in which, a house, in front of the camera, becomes distorted. The only way that this can occur is if there is a lens, between the camera and the house, which is distorting the image. Light coming off the house is being bent and that can only occur if there is an optical interface, between the camera and the house, causing the light to bend. The actual interface can be seen, and its spherical curvature is apparent, in the right image, but it is oscillating in the left image.

Figure 11.1. Still shotsfrom video taken in which the image is distorted by a distorting medium or lens between the camera and house being filmed [1].

The oscillation shows that the interface is not made of anything solid, but that it is created by the atmospheric molecules themselves, which are clearly being manipulated by some outside force. This is therefore an optical interface created out of air molecules, which must be in a very compressed state, in order for the interface to be visble, as it is in the images above. Light is obviously being reflected from a surface which can only be made out of air. This suggests that a standing sound wave is responsible for the creation of this interface. A sound wave is a longuitudinal wave. Its passage through air causes the alternative compression and rarefaction of air molecules.

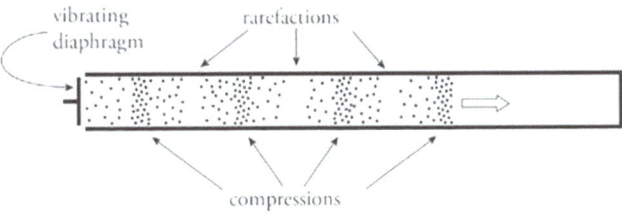

Figure 11.2. A sound wave travelling through a tube containing air molecules causes the air medium to undergo alternative compressions and rarefactions.

A standing wave can be created whenever two waves intersect and interfere. When the two waves are identical and move in opposite directions, a standing wave pattern is created. This causes certain points, in the medium through which the waves travel, to experience alternative constructive and destructive interference. These points are called antinodes. Points where the amplitude is always zero are called nodes. It easier to see how these form when the medium of propagation is a string:

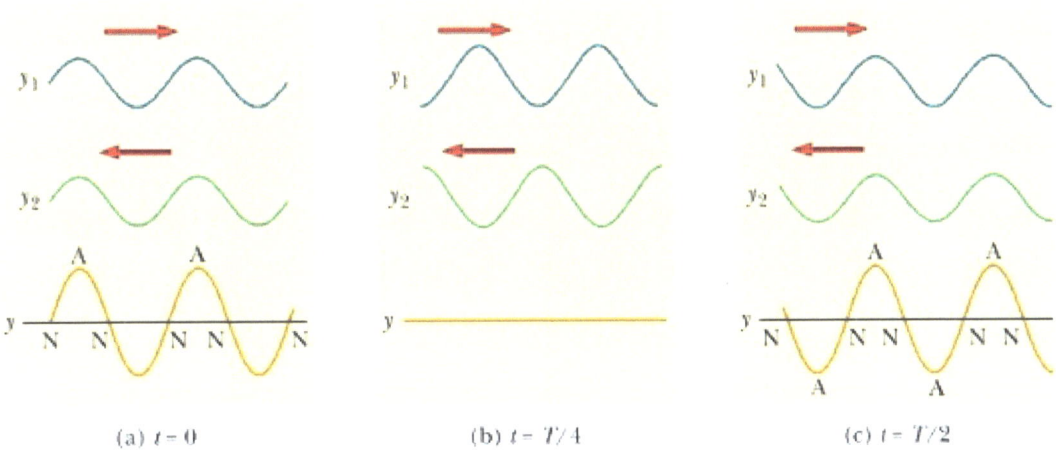

Figure 11.3. A standing wave, in a string, is created when two identical waves, travel across the spring in opposite directions. The points where the string oscillates with maximum amplitude are called antinodes.

The optical interface observed in figure 1 would require very high compression amplitude of the air at the interface. In other words, the air would have to be sharply compressed in that region for the interface to be dense enough to reflect light so that it becomes visible. The air would, however, not be continuously compressed in this region, it would oscillate rapidly, between high compression and high rarefaction, and would thus sharply distort any light passing through it. It is unlikely that such sharp compression and rarefaction can be created out of only two sound wave generators. Quite a few would have to be used from outside an optical interface region and would be difficult to calibrate. It thus seems likely that something else is being used to generate what seems to be a sound wave at the required region within the atmosphere. Since the air has been made conductive by the addition of large amounts of metals, through the dispersion of aerosols, it is possible that the same oscillation in air molecules produced by sound waves, can be produced with the application of electromagnetic waves or oscillating electric fields, in which case, electromagnetic waves would be used to create resonant oscillation of air molecules or sound wave effects. Since seismic waves, are also longitudinal waves, and causing the movement of rock, within, and along, the earth's surface, the same technology, would also be able to create earthquakes, if applied to the earth's crust.

In Article 166: Sun Simulator and lens System [2], I wrote about how the global Sun Simulation system seems to work. This system requires several different devices working together, with at least one Sun Simulator, in orbit, and several operating, within the earth's atmosphere. One of the required components, working in conjunction with the sun and moon simulators, and the holographic projectors, are lenses, within the earth's atmosphere. It has thus now become apparent that these lenses are being created out of air. Interfaces seem to be created out of air through the application of electromagnetic wave technology, aided by the presence of electrically conducting particles such as metals added through chemtrails. These electromagnetic waves seem capable of creating a very dense surface, which will then cause the bending of light or refraction. The fact that they can oscillate rapidly can in addition cause distortion of light. This therefore seems to be a sound wave resonance effect produced through an electromagnetic wave.

In conclusion, a global Sun Simulation System seems to be in operation on planet earth. The system requires that lenses be part of the system. New evidence suggests that these lenses are produced out of air, through the application of electromagnetic technology, which is causing sound wave effects to be generated within the earth's atmosphere. These sound wave effects give rise to optical interfaces, out of alternatively compressed and rarefied air, within the atmosphere.

References:

[1]   Jeff P: https://www.youtube.com/watch?v=uZ1jGtZXIsE&t=165s

[2]   Albers, E. (2018). Article 166: Sun Simulator and lens System.

Chapter 12

Article 214: Global Sun Simulation System and the Dying Sun

In Article 213: Producing lenses out of air in the earth's atmosphere [1], I wrote about the fact that lenses seem to be produced, within the earth's atmosphere, through the application of electromagnetic technology, which then gives rise to resonant sound wave effects, thus producing regions of changing densities of air, which will thus have different, and oscillating refractive indices. These regions will therefore bend light and produce distortion. These lenses are necessary components of the Global Sun Simulation System that seems to be in operation on earth, and which I detailed in Article 166: Sun simulation and lens system [2]. This system appears to have several different components, some will be in orbit, and others will operate within the atmosphere. At least one simulator appears to be in orbit and this simulator appears to be producing most of the surface's illumination around the midday position. However, it is not likely that it can simulate the Sun close, to the sunset, and the sunrise positions. Sun simulators, within the atmosphere, would be required to simulate the sun, low on the horizon. In addition, the Sun simulator in orbit is not likely to be able to adequately illuminate these peripheral regions.

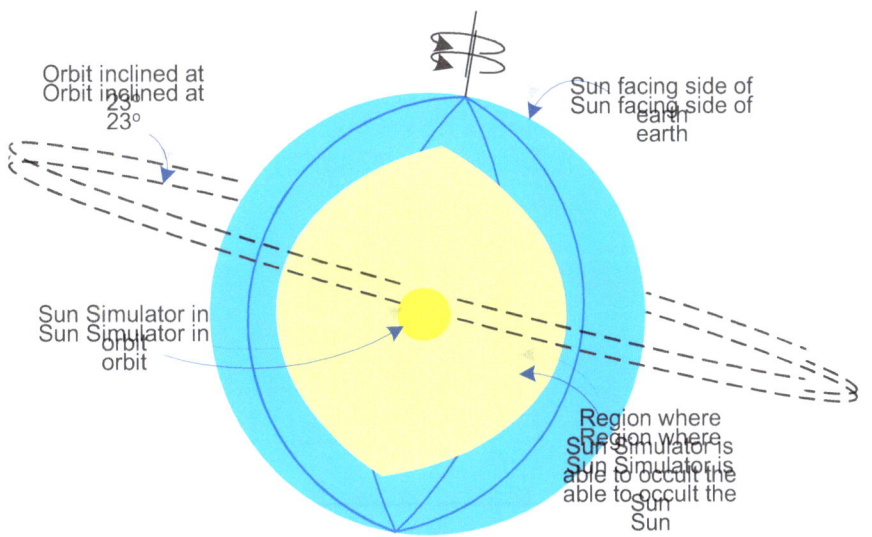

Figure 12.1: The blue region represents the Sun facing side of planet Earth. The Sun Simulator maintains an orbital altitude (12 500 mi) and corresponding speed, such that it is always over a position, on the surface of the Earth, above the Equator, which corresponds to midday. The orbital Sun Simulator is able to illuminate the beige region but not the region beyond that, and thus simulators, operating within the atmosphere, would be required to illuminate the blue region [2].

Evidence that atmospheric simulators are indeed operating, within the earth's atmosphere, arises out of the many photographs showing diverging light beams, coming through holes, in clouds. And, now, there is additional evidence coming from a recent Thirdphaseofthemoon YouTube video, which shows a light source, within thick cloud, filmed from above the cloud cover.

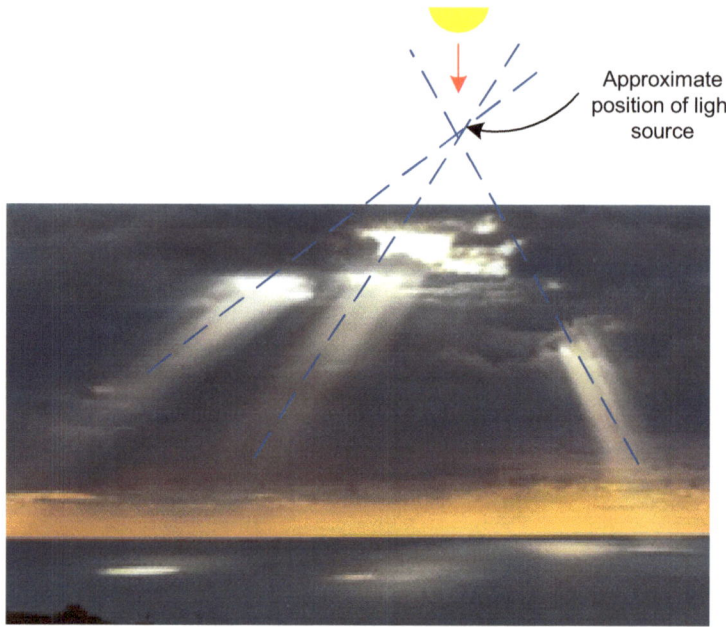

Figure 12.2. Diverging light rays produced by large holes, in a thick cloud, indicate that the light source is not at infinity, but close. Extending the rays of light, coming through holes in the cloud, in this photograph, backwards, allows the position of the light source to be determined. This position is not far above the clouds. This means that the source is artificial, i.e. a Sun Simulator. This light source is sufficiently far away not to move, if the observer moves a short distance, and it is thus, only, when its light is interrupted, by a cloud, that its proximity, and position, within the atmosphere, becomes obvious.

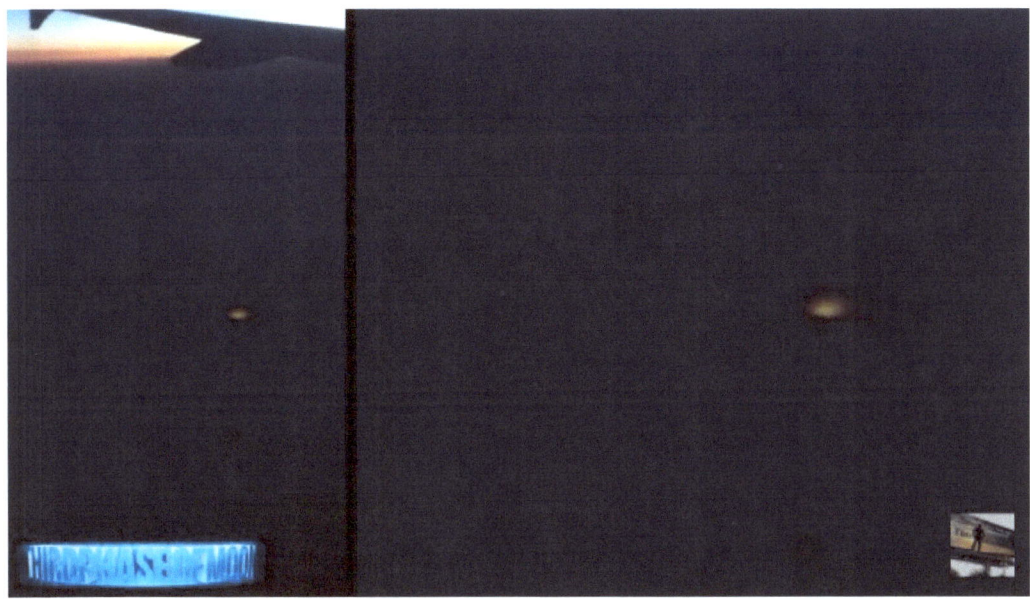

Figure 12.3. Screenshot from video footage presented in a thirdphaseofthemoon video [3] showing a light source within thick cloud as seen from an airplane. The lighting suggests that it is close to sunset, in this region of the earth, and this is therefore the time, when simulators would be required to be in operation, within the earth's atmosphere.

The fact that the Sun now looks white and bluish, and never yellow, within the earth's atmosphere, as the real Sun would, is evidence that we are not seeing the real Sun, from within the earth's atmosphere. The earth's atmosphere scatters blue light, out of the white light, emitted by the real sun. This is the reason why the sky used to look blue, and the sun, yellow. Other evidence that what is now operating, in the earth's atmosphere, is not the Sun, is the obvious multiple beams emanating from the central light source, shown below, as well as the multicolored additional light sources, associated with it, and congregated around it. In addition, there seem to be some disk shaped objects, with what seems to be holes in them, forming an array. The main light source has 6 double beams, emanating from it, showing that it has a hexagonal symmetry, and cannot therefore be a natural object.

Figure 12.4: Hexagonal (6 sided) Sun simulation device surrounded by multicolored additional light sources and multiple other disk shaped devices, some with what seems to be holes in them, in an array type of pattern. These are all clearly artificial and not what our real Sun could possibly be like:

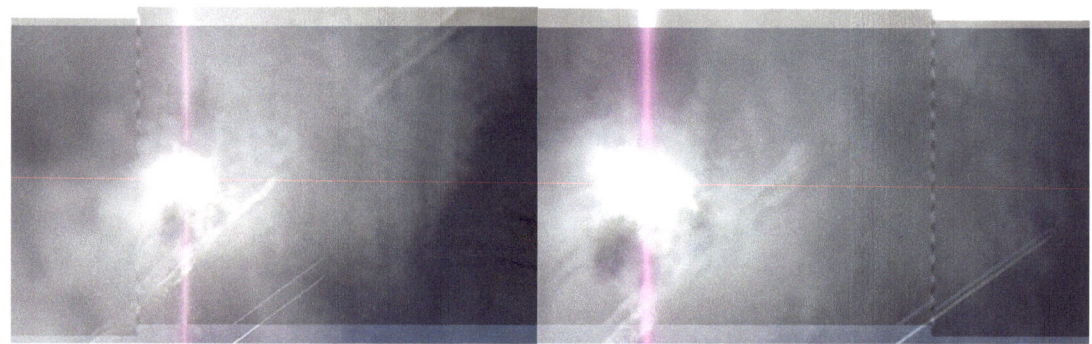

Figure 12.5: Photographs taken by Scott C'one: The pink laser beam through a Sun Simulator suggests that laser beams are used to take power to the devices. The laser beam was in place for a limited period of time as photographs taken a few minutes later did not show it [4]:

In Article 166: Sun Simulation and lens system [2], I also mentioned that the Sun Simulators, whether operating from orbital altitude, or operating within the atmosphere, are likely to be getting power from the earth's ionosphere, and that they may be connecting to it through laser beams. Now, a recent photograph, by Scott C'one, from Planet X News YouTube channel [4], shows a magenta (pink) colored laser beam, going through an operating Sun Simulator, suggesting that powerful laser beams are, indeed, being used to transmit power to the devices.

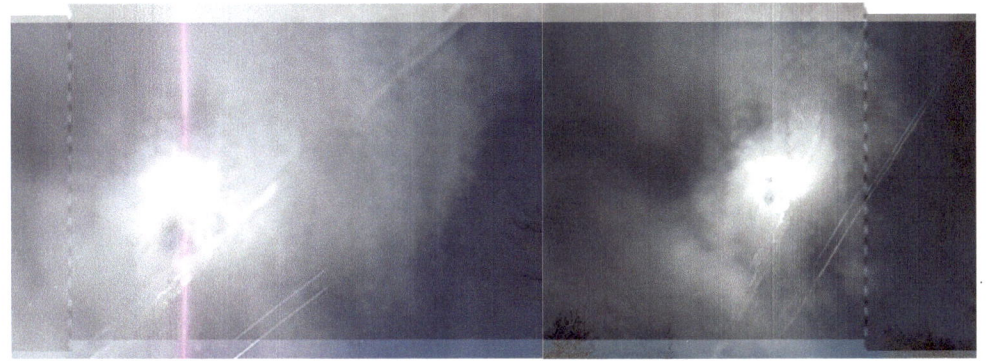

Figure 12.6: Left: A pink laser beam through Sun simulator. Right: Photograph taken of the same Sun Simulator, after the laser beam had shut off, indicates that its presence, in the left and above, photographs, was not an optical effect produced by the camera [4].

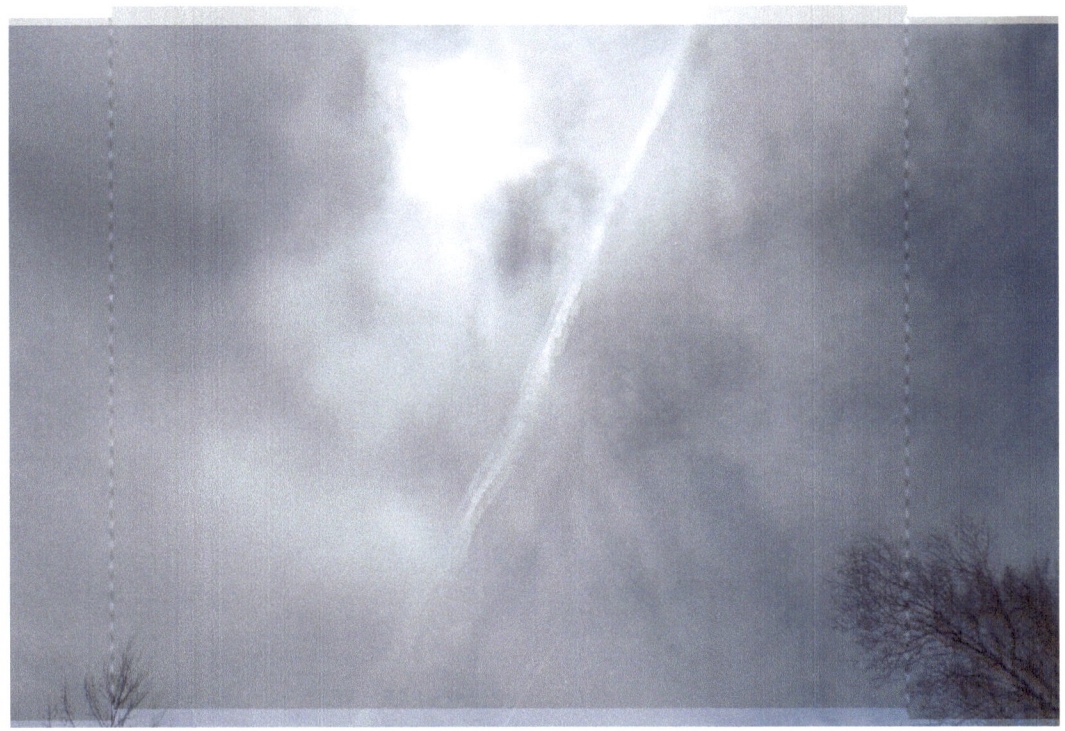

Figure 12.7: Photograph of an artificial light source and a dispersing chemtrail cloud. The dispersing chemtrail seems to have created a shadow of itself, on the cloud above it. This is only possible if there are light sources below the chemtrail cloud shinning upwards.

In the above photograph the large chemtrail which is starting to disperse is producing a shadow on the cloud above it. This can only be produced if a light source is illuminating the dispersing chemtrail cloud from below. This suggests that the earth's atmosphere is being illuminated by artificial light source that are shinning upwards. The only logical region for doing this is to illuminate the atmosphere sufficiently so that the sky looks blue, from the ground. If this is necessary, it suggests that the Sun is now so dim that the sky would not look blue, if this illumination was not supplementing, the illumination that the Sun simulators can provide.

Figure 12.8. Another photograph which indicates the long dispersing chemtrail is producing a shadow on cloud above it. This photograph was taken before the one shown in figure 7 and the dispersing chemtrail, as well as its shadow, are not as wide as in figure 7.

The reason for the use of the global simulation system, seems to be to hide the fact that the Solar System has been invaded by a System of Stellar Cores (see Article 116: Planet X Objects: unbelievable evidence and size) [5], which are absorbing energy from the Sun, and causing it to continuously weaken. Some of these objects seem to also be in orbit, around the earth, and to be emitting light (see Article 188: What is causing the ocean to recede all over the world) [6]. Current SDO images suggest that the Sun is now much weaker, than it was years ago (see Article 195: Stellar Cores and the dying Sun) [7]. In addition, the same SDO images indicate that the Sun is going completely dark, periodically, and that this seems to be happening more frequently (see Article 211: Planet X and Sun interaction: Sun goes dark on April 18th 2018) [8].

Nowadays, whenever the Sun is close to the midday position, there seems to be a halo around it, and the sky, within the halo, looks grey instead of blue. The sky is supposed to look lighter blue, the closer it is to the Sun's position, but the opposite is occurring. The sky looks light blue close to the horizon, and looks darker blue, as we move toward the Sun, when it is high in the sky, and then it turns grey, within the halo, which is usually present around the Sun. The grey color may be due to the fact that the Sun simulator, in orbit, is designed to work like a plasma ball. In other words, the light source is contained within a transparent sphere. But the darker color of the sky, beyond the halo, suggests that the Sun is now extremely dim, and the Sun simulator used at the midday position is not illuminating the atmosphere sufficiently, for the sky to appear as light blue, as when the real Sun was still emitting a normal amount of light. The sky, closer to the horizon, will most likely look lighter blue, because it is being illuminated by numerous, in the atmosphere, simulators, and we are looking through more of the atmosphere, when we look at the sky close to the horizon.

Figure 12.9. Photograph showing the Sun looking extremely white with a halo around it. Even the strangely shaped chemtrail clouds cannot hide the fact that the sky is grey within the halo and that the sky looks darker than is normal around even outside the halo. Chemtrails seem to be heavily used in an attempt to hide this.

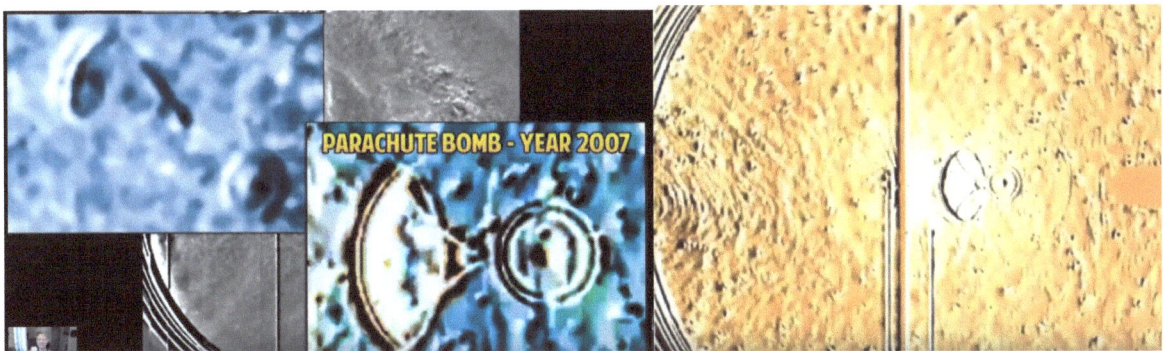

Figure 12.10. Device given the name 'Parachute Bomb', which appeared in SECCHI images, from an article I wrote about it, in 2016. The circular part of the device always looks circular, no matter the orientation of the parachute part of it, thus indicating that it is spherical. It is also transparent as space could be seen through it. This object is most likely a large plasma ball type device and it is also very likely to be the Sun simulator we often see from the earth's surface close to the midday position. This

image comes from a Chris Potter video and is from a time when he still seemed to believe in what he was doing.

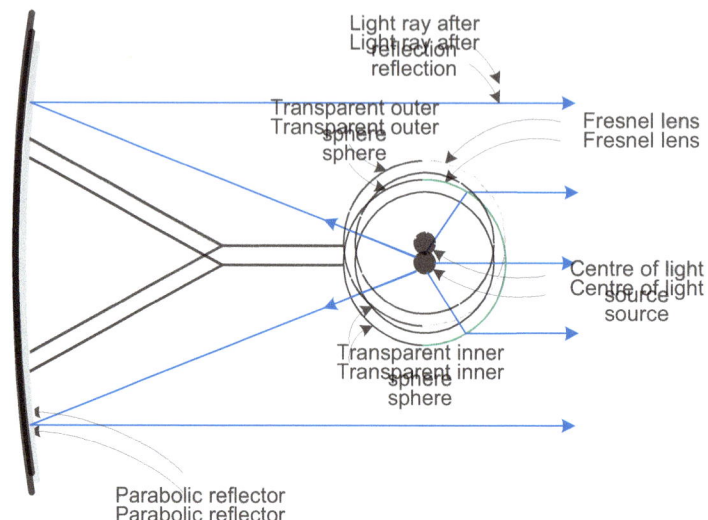

Figure 12.11: Parachute Bomb (Sun simulator) design: The blue lines are light rays that emanate from the center of the light source, which is at the center of 2 transparent spheres. The outer sphere has a lighthouse (Fresnel) lens on the forward facing half of the sphere. The parabolic reflector reflects all rays forward so that they end up parallel to each other.

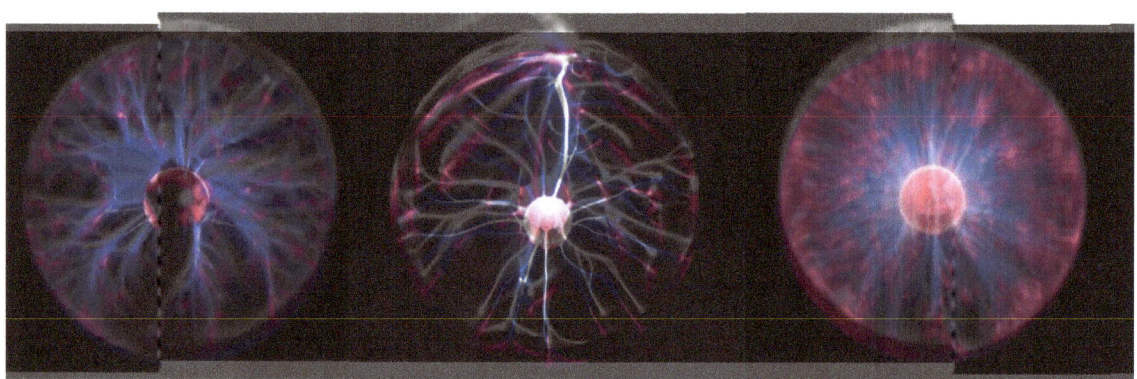

Figure 12.12: Plasma lamps that create light in much the same way as the light source in the Sun simulator Parachute bomb would.

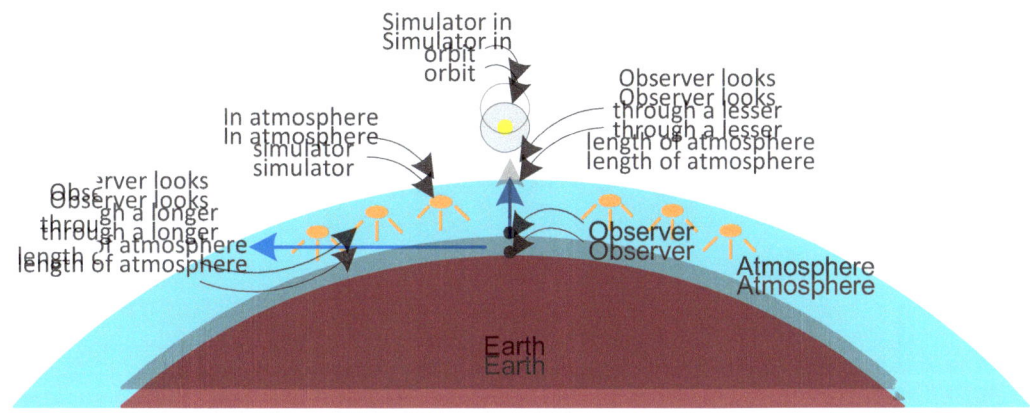

Figure 12.13: When looking at the sky close to the horizon, the observer looks through a longer length of atmosphere which is being illuminated by several in the atmosphere simulators and thus the sky seems to be a paler shade of blue. The in orbit simulator is not able to compensate for the fact that the Sun is now too dim to properly illuminate the atmosphere and thus the atmosphere looks dark blue when observer looks upwards.

The fact that the atmosphere is now also being illuminated, from below, also indicates that an attempt is being made to make illuminate the atmosphere further, in order to make it lighter blue, and perhaps also in an attempt to cause reflection, off the chemtrail clouds, below the simulator in orbit, so that the dark grey region, inside the halo, is not as noticeable. All these are likely a sign that the Sun is much dimmer, than even the SDO data suggests. Since it is unlikely that we are seeing the real Sun, from the earth's surface, at any time, it is possible that the SDO data is being manipulated, to give the impression that the Sun is in a better state, than it really is. It is even possible that the Sun is now completely dark, most of the time.

In conclusion, there is substantial evidence that a global Sun Simulation System is in operation, on planet earth. Some of the components of this system, operate from an orbital position, and some operate, from within the atmosphere. Powerful pink colored laser light beams seem to be the method through which power is transmitted to these devices. The fact that the earth's atmosphere is being illuminated, from below, with the likely aim of producing a blue sky, and that the sky looks darker close to the sun simulator, when it is high in the sky, or close to the midday position, suggests that the Sun may be in a worse state, than even the current available SDO data would suggest.

References:

[1] Albers, C. (2018). Article 213: Producing lenses out of air in the earth's atmosphere.
[2] Albers, C. (2018). Article 166: Sun Simulator and lens System.
[3] Thirdphaseofthmoon: https://www.youtube.com/watch?v=rXIrhyUhgMM&t=0s
[4] Scott C'one, Planet X News: https://www.youtube.com/watch?v=vksQZ4ZETkw
[5] Albers, C. (2017). Article 116: Planet X Objects: unbelievable evidence and size.
[6] Albers, C. (2017). Article 188: What is causing the ocean to recede all over the world.
[7] Albers, C. (2018). Article 195: Stellar Cores and the dying Sun.

[8]   Albers, C. (2018). Article 211: Planet X and Sun interaction: Sun goes dark on April 18th 2018.

Chapter 13

Article 204: Harmful UVC radiation reaching earth's surface indicates source within atmosphere

Recent published research on the amount of ultraviolet radiation, entering the earth's atmosphere, shows that all wavelengths in the ultraviolet range are reaching the surface of the earth. The research collaborators were Marvin Herndon, from the Transdyne Corporation, Raymond Hoisington, from iRay SpectraMetrics, and Mark Whiteside, from the Florida Department of Health in Monroe County. The measurements were carried out in June of 2017 and January of 2018 and are summarized in the graph below [1].

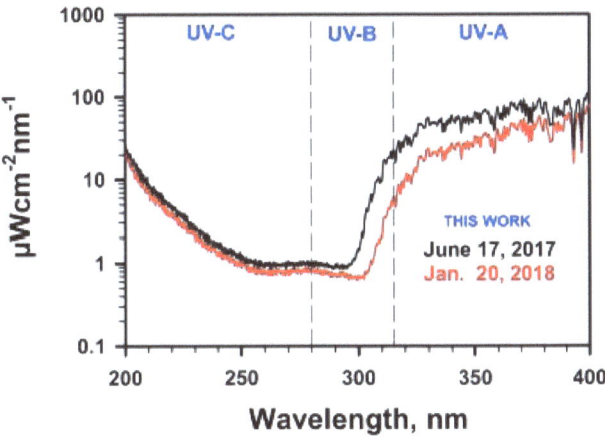

Figure 13.1. Measurement of spectral irradiance per unit wavelength, at the earth's surface, for the ultraviolet spectrum range [1]. Large amounts of the highest energy UV C radiation is shown to be reaching the earth's surface.

Spectral irradiance is a measure of light intensity (power per unit surface area) and gives a measure of the amount (power which is energy per second) of light impacting a particular surface. The measurements can be made at different altitudes, including above the earth's atmosphere, but in this case, they were made very close to sea level, i.e. at an altitude of 56 m, above sea level.

The ultraviolet spectrum is divided into 3 range groups. Photons with a wavelength between 315 and 400 nm fall in the UV A range, which is the least energetic ultraviolet range. The middle range group is UV B. Photons with wavelengths between 280 and 315 nm fall in this range. Ultraviolet light in both these wavelength ranges are beneficial to living organisms. Ultraviolet light is helpful in certain conditions where the body is attacked by bacteria and viruses, and ultraviolet light therapy has been successfully used to treat psoriasis, eczema and jaundice. Exposure to ultraviolet light stimulates the skin to produce vitamin D. Vitamin D facilitates the absorption of calcium and phosphorus, from food, and is thus crucial for skeletal development and blood cell formation [2].

However, radiation in the UV C range, from 100 to 280 nm, is the most energetic and is part of the radiation range, which is able to ionize atoms, and is thus called ionizing radiation. Ionizing radiation is damaging to living cells. Apparently DNA has a resonant frequency at 270 nm, so UV rays at that frequency cause an intense vibration that destroys DNA. Thus, when we are exposed to UV C radiation, rampant DNA damage occurs in exposed skin and in the eye. UV C radiation is also damaging to plants, as it inhibits photosynthesis, and destroys beneficial symbiotic soil organisms, which help plants absorb certain minerals. Thus, UV C radiation can destroy crops and thus the food supply we are dependent on for our survival.

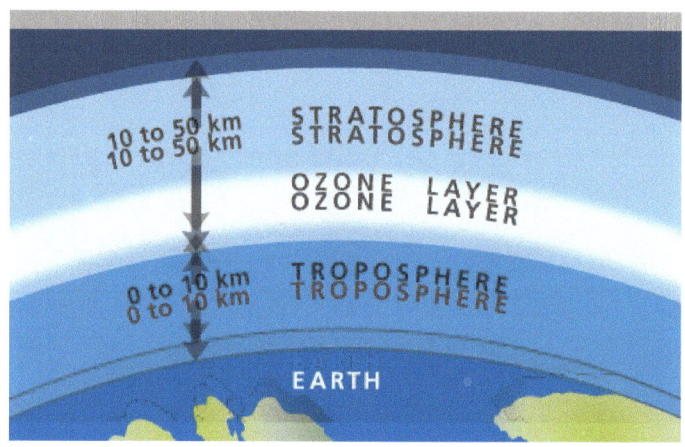

Figure 13.2. The ozone layer in the earth's atmosphere filters UV C radiation, which is damaging to all living organisms

Now, UVC radiation is believed to be 97 to 99% absorbed in the earth's atmospheric layer called the ozone layer. Living organisms have repair mechanisms that allow them to cope with such small amounts. But when too much reaches the surface the repair mechanisms are overwhelmed and disease and death can result. Now, there has been some evidence that the earth's ozone layer has been undergoing depletion, for quite some time, now, and that this may be the root cause behind the destruction of Australia's coral reefs. Since the spraying of aerosols (chemtrails) is known to deplete the ozone layer, and this has now been done, on a large scale, for about 10 years, it would not be surprising if the ozone layer has now been depleted to the point that UV C radiation is reaching the earth's surface, in large amounts (see Article 32: The purpose and effects of chemtrails) [3].

Hendon et al. concluded in their paper that the common belief that the earth's atmosphere absorbs all UV C radiation is erroneous, and that, in fact, it does not; it, instead, allows UVC radiation to reach the surface. I do not agree with that conclusion. I believe that the more logical conclusion is that there has been a dramatic change, in the earth's atmosphere, between 2002 and 2017. I base that conclusion on the fact that measurements carried out by Brian Diffey, in 2002, clearly showed that UVC radiation was not detected at the earth's surface, then. Hendon et al. superimposed Diffey's measurements, from 2002, on their own results, and this graph appears in figure 2 below.

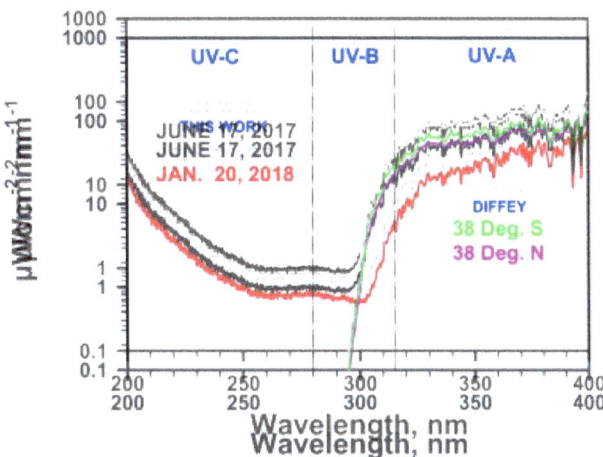

Figure 13.3: Measurements of ultraviolet spectral irradiance, done by Diffey, in 2002 [4], superimposed on the measurements done by Hendon et al. This seems to show that something seems to have significantly changed in the earth's atmosphere between 2002 and 2017 so that, now, a large amount of UVC radiation is getting to the earth's surface.

Now, besides the fact that the earth's ozone layer has gone through dramatic depletion since 2002 due to the large scale spraying of aerosols, in the earth's atmosphere, it is also possible that the sun simulators, which I have shown are operating from within the earth's atmosphere (see Article 165: Sun Simulator: irrefutable evidence) [5]. In Article 166: Sun Simulator and lens system [6] I explain how the global sun simulation system works. Although, it is likely that at least one of the sun simulators is operating from an orbital position, where it has an orbital period of 24 hours, most of the simulators, in operation, seem to be doing so inside the earth's atmosphere, and moving relative to the earth's surface at a speed of about 1000 mi/h.

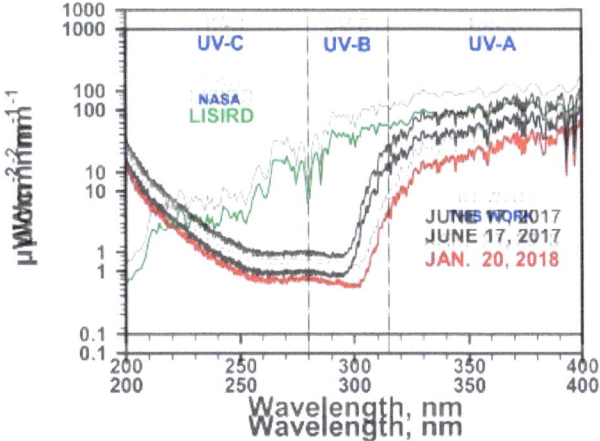

Figure 13.4: Hendon et al. spectral irradiation results superimposed on NASA's LISIRD results, obtained above the earth's atmosphere. This shows that an extra source of UVC radiation between 200 and 220 nm has to be responsible for the results obtained by Hendon et al.

Hendon et al. measured more UVC radiation between the 200 and 220 nm range, than is measured above the atmosphere, and thus before the ozone layer, can remove any of this radiation. This shows that there has to be a source for the radiation below ozone layer altitude. This strongly suggests that, indeed, the Sun simulators operating inside the earth's atmosphere, are emitting UV C radiation and are therefore harmful to living organisms, on the earth's surface. It would be interesting to see if measurements at other wavelengths, such as the x-ray range, would also yield positive results.

In conclusion, damaging UVC radiation is now reaching the earth's surface. The reason is likely to, at least in part, be due to depletion of the earth's ozone layer. However, the fact that more UVC radiation is measured at the earth's surface, than has been measured above the atmosphere, suggests that a large part of the UV C radiation, reaching the earth's surface, cannot be coming from the Sun, and is very likely to be coming from sources, below the ozone layer and therefore, most likely from the Sun simulation devices in operation, within the earth's atmosphere.

References:

[1] Herndon, J. M., Hoisington, R. and Whiteside, M. (2018). Journal of Geography, Environment and Earth Science International 14(20) pp.1 -11.
[2] http://www.who.int/uv/faq/uvhealtfac/en/index1.html
[3] Albers, C. (2017). Article 32: The purpose and effects of chemtrails.
[4] Diffey, B. (2002). Sources and measurement of ultraviolet radiation. Methods 28(1) pp. 4-13.
[5] Albers, C. (2018). Article 165: Sun Simulator: irrefutable evidence.
[6] Albers, C. (2018). Article Article 166: Sun Simulator and lens system.

Chapter 14

Article 147: The chemtrail and iridescent cloud connection

Figure 1 below shows a photograph of an iridescent cloud. I believe that the official explanation that the iridescent effect is caused by diffraction is correct, but it is essentially thin film diffraction, and ice crystals or water droplets do not form thin films on water. If they did, we would have always had iridescent clouds. However, they have not always been around. Iridescent clouds are a new phenomenon that is becoming more prevalent.

Figure 14.1. Iridescent clouds have multiple bands of different colors caused by interference of light or diffraction.

Different gases emit different colors of light, when ionized. But, the alternative bands of different colors, seen in iridescent clouds, cannot be produced by the ionization of different gases because gases, get thoroughly mixed by winds, in the atmosphere, and cannot therefore, occur as separate bands. In other words, it is not possible to get a band of oxygen next to a band of argon gas.

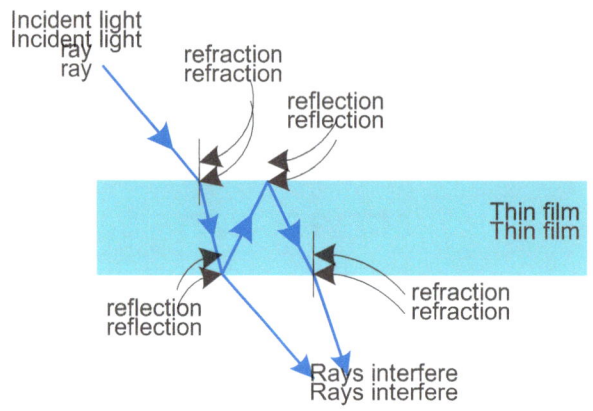

Figure 14.2: Thin film diffraction: a light ray interferes with itself.

Thin film interference is caused by rays of light being refracted and reflected, at the boundaries of a thin film, which is usually of about the same thickness as the wavelength of the light (550 nm for green light), passing through it. Figure 2 above shows a ray of light being refracted at the boundary between air and water, above the thin film. A part of the ray is transmitted, and so only undergoes one more refraction, as it exits the bottom, of the thin film. But the other portion of the ray, does not exit and is instead reflected at the bottom boundary, then reflected at the top boundary, and finally it is again refracted as it exits, at the bottom of the thin film. This results in one ray being split into two rays, which then interfere with each other. In other words, if when the two rays reach the eye, of the observer, they have the same amplitude (for example: both at maximum amplitude or crest position) then they interfere constructively and a light band is seen, but if they have opposite amplitudes (for example: one has a maximum amplitude and the other a minimum or a crest with a trough) they interfere destructively and a dark band is seen. Now, visible light is made up a range of wavelengths, so we end up with different wavelengths interfering constructively and destructively, from different points on the thin film, resulting in a diffraction pattern, as shown in figure 3 below.

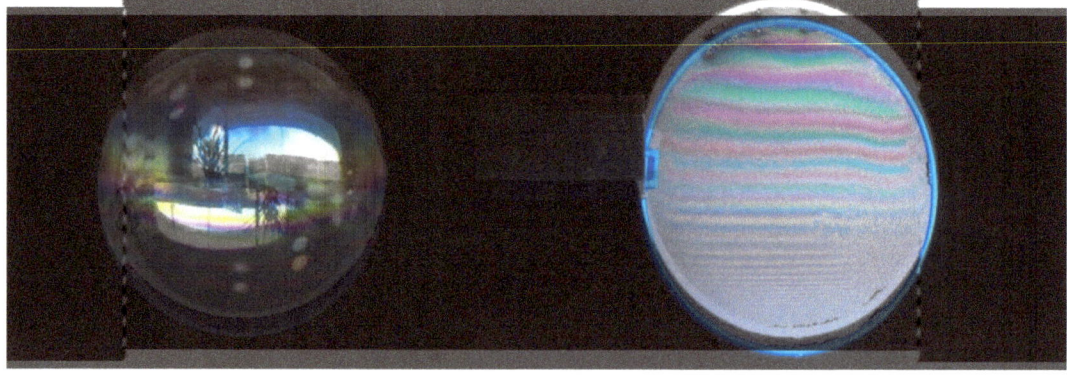

Figure 14.3: Diffraction patterns produced by diffraction of light, in film films.

The diffraction patterns, shown in figure 3, are for light being reflected off a thin film and interfering with itself. The pattern, we see in the iridescent clouds, though, is produced by light that is transmitted through the thin film, but the pattern, will look exactly the same.

Now, in order to have thin film interference or diffraction in the sky, we have to have thin films in the sky. Thin films are widely used in industry such as in antireflective coatings on lenses, solar cells, magnetic recording media, and in the pharmaceutical industry, as a form of drug delivery.

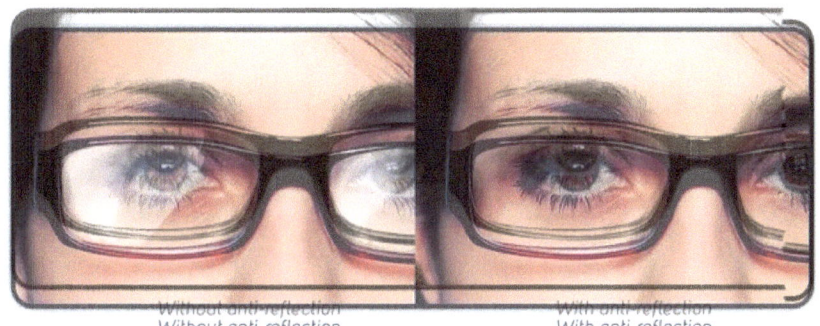

Figure 14.4: Thin films are used as anti-reflection coatings on glasses. They cause destructive interference of light reflecting off the lens so that it does not reach the eye.

Different compounds are well suited to the making of thin films, namely: silicon compounds, such as silicon dioxide and silicon nitride: carbon compounds in the form of carbon fibers, carbon nanofibers and carbon nanotubes; and in particular, fluorocarbons. Of particular interest are fluoropolymers, the best known of which goes by the brand name: Teflon. These are of particular interest because they are available as hydrophilic formulations. In other words, they can be mixed with water and spread through aerosols.

Figure 14.5: Straight line clouds are not normal. These are seeded by aerosols spread by jets in the upper atmosphere. These clouds spread out from the thin white line of aerosol left by the plane as it passes overhead

Now, research has shown that fluoride compounds are neurotoxic, producing lower IQs and learning disabilities, in children, and producing difficulties in following social norms, leading to criminal behavior. During the Second World War, both the Germans and the Soviets added sodium fluoride, to the drinking water of prisoners of war, in order to make them stupid and docile. Is it possible that since people increasingly refuse to drink water, with fluoride in it, that the powers that be have found another way to force it down our throats? This time, through the air we breathe? And at the same time increase the haze in the atmosphere so that objects in our skies remain well hidden?

Figure 14.6. Jet planes spread aerosols loaded with toxic chemicals which include barium, strontium 90, aluminum, cadmium, zinc, viruses and chaff, which are poisoning plants, animals and humans.

In conclusion, the phenomenon of iridescent clouds seems to be produced by thin film diffraction, due to the presence of thin films in our atmosphere. And, these thin films are very likely to be made of hydrophilic formulations of fluoropolymers which are added to the chemtrail mix that has been used for many years now.

Chapter 15

Article 32: The purpose and effects of chemtrails

The fact that our skies are being sprayed with chemicals is well known to many. Anyone, who lives in an industrialized nation, will look up into the sky and see planes flying at high altitude and leaving a double white trail behind them, almost every day, as shown in figure 1 below. The proper name for what they are doing is geoengineering. This means that an attempt is being made through this spraying of chemicals, in the upper atmosphere, to change the earth's climate. Reports of this spraying of chemicals, in the upper atmosphere, date back to 1977, but it does not seem to have been done on a large scale until 1997. This means that a concerted effort to change the earth's atmosphere has now been carried out for 20 years.

Figure 15.1. A plane sprays a deadly cocktails of chemicals, in the earth's upper atmosphere.

The chemicals the planes are being used to spray, in the earth's atmosphere, appear to be mostly aluminum and barium, but they also include other elements and compounds such as CHAFF, which is fiberglass coated with nanoparticle aluminum, radioactive thorium, cadmium, chromium, nickel, desiccated blood, mold spores, yellow fungal mycotoxins, ethylene dibromide and polymer fibers such as fluoropolymers. Barium is known to induce heart disease and aluminum to induce brain disease. Aluminum is a known cause of dementia, which seems to be occurring at a younger and younger age. Symptoms of dementia include: emotional outbursts, paranoia, forgetfulness and memory loss, speech incoherence, irritability, diminished alertness, changes in personality, and poor judgment.

Fluoropolymers contain fluoride. The best known fluoropolymer goes by the brand name: Teflon. These are available as hydrophilic formulations and can therefore be mixed with water and spread through aerosols. Once in the atmosphere these compounds can form thin films and give rise to iridescent clouds, also known as rainbow clouds, as a result of thin film diffraction.

Now, research has shown that fluoride compounds are neurotoxic, producing lower IQs and learning disabilities, in children, and producing difficulties in following social norms, leading to criminal behavior. During the Second World War, both the Germans and the Soviets added sodium fluoride, to the drinking water of prisoners of war, in order to make them stupid and docile. Is it possible that since people, increasingly, refuse to drink water, with fluoride in it, that the powers that be have found another way to force it down our throats? This time, through the air we breathe? And at the same time, increase the haze in the atmosphere, so that objects in our skies remain well hidden?

Also, the large concentration of these particles, in the air we breathe, affects lung function, and leads to increased incidence of asthma, and other respiratory diseases. Radioactive thorium and most of the other components are carcinogenic. In fact, the toxic brew seems to be designed to make all living organisms, on earth, overloaded with toxicity and diseased. Is this an effort to kill most of the earth's population? Well, a poison is a substance that when ingested is able to cause illness or death of a living organism. Since poison is being deliberately administered to most of the world's population, I would say that yes, an effort seems to have been made and continues to be made to kill as many living organisms, including human beings, as possible, on earth.

Since the spraying is usually concentrated around the Sun, and since there are several objects close to the Sun, part of the reason for geoengineering seems to be the hiding, from the earth's human population, of the objects orbiting and draining our Sun of energy.

Figure 15.2: Geoengineering grid pattern seems to purposely block our view, of the Sun, and the area of sky surrounding it. This suggested that at least one of its aims is to hide the presence of objects that are surrounding the Sun.

But one of the most devastating effects of geoengineering seems to be the destruction of the earth's ozone layer, which protects every living organism, on earth, from the destructive effects of the Sun's most energetic ultraviolet radiation, known as UV C rays. UV C rays have a wavelength range between 100 and 280 nm. Apparently DNA has a resonant frequency at 270 nm, so UV rays at that frequency cause an intense vibration that destroys DNA. Thus, when we are exposed to UV C radiation, rampant DNA damage occurs in exposed skin and in the eye.

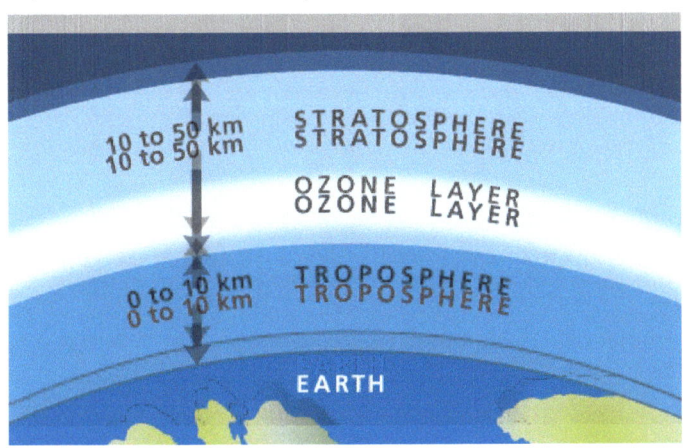

Figure 15.3: The ozone layer in the earth's atmosphere filters UV C radiation, which is damaging to all living organisms.

The ozone layer is a filter, which absorbs UV C radiation, but it has never been a perfect filter, as it only absorbs 97 to 99 percent of the radiation. In other words, 1 to 3 percent has always been able to get through the ozone layer, to the earth's surface. But since organisms have DNA repair mechanisms, they are able to deal with the resulting small amount of DNA damage.

However, geoengineering has resulted in such a depletion of ozone, in the ozone layer, that large amounts of UV C radiation is now getting to the earth's surface, and causing untold damage. Wherever this is happening, the sun will seem to be a lot hotter, and sunburn will also happen a lot quicker. The DNA damage will also cause tiredness, as the body assigns more resources to DNA repair mechanisms, in an effort to repair damaged DNA, but if it is unable to keep up with the damage, cancer and immune system failure may be the result, leading to death.

Ozone is a molecule made up of 3 oxygen atoms, and the ozone layer is a layer, in the earth's atmosphere, that contains a much higher concentration of ozone (10 parts per million) than the rest of the earth's atmosphere, which has an average of only 0.3 parts per million. Ozone (O_3) is formed when the Sun's UV C radiation is absorbed by a normal oxygen atom (O_2), which splits the molecule into 2 oxygen atoms (O) in the process. Then the separated oxygen atoms combine with oxygen molecules to produce ozone molecules. This is illustrated in figure 4 below. However, this reaction is slow. In other words, ozone is produced slowly through this reaction, and therefore not much UV C radiation is absorbed through it.

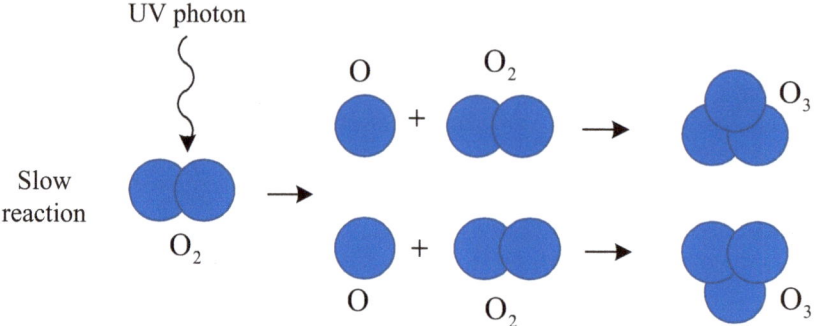

Figure 15.4. Illustration of the slow reaction, through which, ozone is created in the upper atmosphere.

But when a UV C photon encounters an ozone molecule (O_3), it splits one oxygen atom from the ozone molecule, which results in one normal oxygen molecule (O_2) and one oxygen atom (O). The UV C photon is absorbed in the process. Then the one oxygen atom (O) combines with an oxygen molecule (O_2) to form another ozone (O_3) molecule. Figure 5 illustrates the reaction process. This reaction happens very fast and so a lot of UV C photons are absorbed, as long as there is plenty of ozone in the ozone layer.

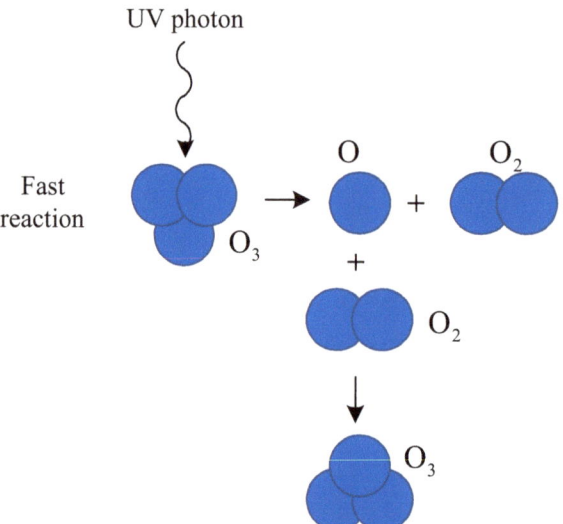

Figure 15.5. Illustration of the fast reaction, through which, UV C photons are absorbed, without any depletion of ozone, in the ozone layer.

However, if there are small particles, like aluminum or Sulphur, in the upper atmosphere, when the ozone molecule (O_3) splits into an oxygen molecule (O_2) and a single oxygen atom (O), the single oxygen (O) then combines with another element instead of an oxygen molecule (O_2) so that the ozone molecule that should then form, does not. Thus, the amount of ozone decreases. In other words, if there is Sulphur in the ozone layer, which mainly comes from volcanic eruptions, or nanoparticle sized aluminum , which comes from geoengineering sprays, once the UV C rays splits ozone molecules into normal oxygen molecules, and single oxygen molecules, the single oxygen molecules then combine with either Sulphur or aluminum instead of with oxygen molecules. This leads to the formation of aluminum oxide

(Al_2O_3) and Sulphur dioxide (SO_2) instead of ozone (O_3). So, the ozone concentration in the ozone layer drops, as illustrated in figure 6 below.

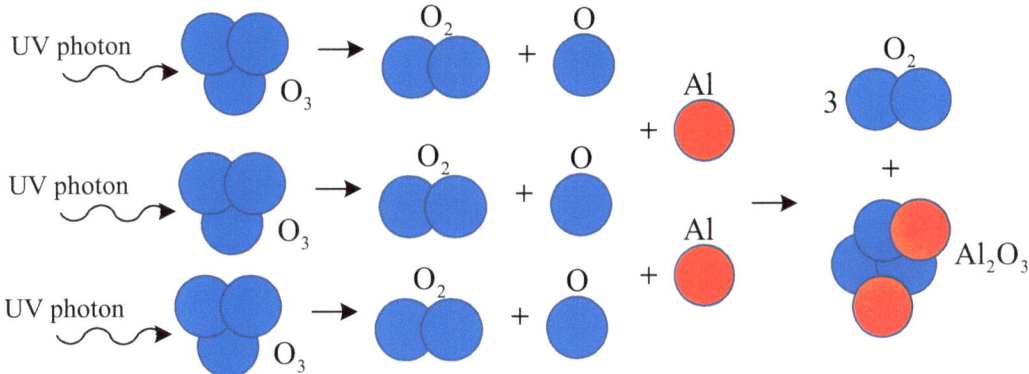

Figure 15.6. Illustration of how ozone is depleted, if aluminum is present in the upper atmosphere.

Now, volcanic eruptions may place small sized particles in the atmosphere, which may lead to a temporary depletion of ozone, but geoengineering is continuously placing large amounts of these particles, in the upper atmosphere, and have been doing so, for 20 years. In fact, the geoengineering effort seems to be intensifying and smaller and smaller particles seem to be used, which leads to the whole sky seeming to be covered in a pale haze. This type of spraying may not be as noticeable to the world's population, but it is even more damaging because it has to be produced by even smaller sized particles, and the smaller the particles, the more they will be able to deplete ozone, through the mechanism described above. This is because it becomes even easier for single oxygen atoms to combine with the elements, these particles are made of.

So, what is the result of this destruction of the ozone layer? It produces ozone holes over certain regions of the world. These are areas, where there is just about no ozone in the upper atmosphere and the UVC radiation, comes right down to the earth's surface, and causes untold destruction to living organisms. The ozone hole may not necessarily be due to direct spraying over that region but sometimes the ozone hole is brought in by trade winds. There have been reports of such an ozone hole occurring over Australia for many years now. There have also been reports of large numbers of the population, in certain cities in Australia, feeling very tired over the last summer and having problems with vegetables not growing, or plants not producing fruit. These are typical symptoms of a worsening condition, over Australia, in regard to ozone depletion. The same seems to be occurring over the US, as people often report feeling a burning sensation, on their face, from any amount of sun exposure. This is another symptom that is to be expected, from ozone depletion, or an ozone hole, over the area.

But ozone depletion would also lead to the destruction of sensitive organisms, such as coral and plankton. A BBC report from November 28th 2016 reported on the death of Australia's extensive coral reefs. There was a 67 % destruction, in the reefs to the north of the continent. This is a huge level of destruction, which is usually attributed to warming oceans but the real reason is ozone depletion, which allows UV C radiation to arrive at the surface of the planet leading to DNA damage and destruction of the coral.

Plankton supports the ocean's ecosystem, and if it is depleted as a result of UV C radiation, due to ozone depletion, in the ozone layer, and it affects all sea life. The plankton is eaten by small fish, and the small fish by bigger fish and thus its availability can affect the whole ocean food chain. The plant part of plankton, phytoplankton, also absorbs CO_2 and gives off oxygen, and is therefore of paramount importance to the earth's entire ecosystem. Phytoplankton produces 50% of the earth's oxygen, through this cycle. Ozone depletion therefore places all life, on earth, in danger. But green plants, on land, are also in danger and UV C radiation will also destroy unprotected food crops. This will in turn endanger human life, since we are dependent on food crops for our survival.

What can we do to protect ourselves, in the meantime? We can try to stay indoors, as much as possible, and if we have to go out, we can try to cover every possible portion of skin to minimize exposure to UV C radiation. If we do this, it may also be a good idea to take some vitamin d, as the body will no longer be making any through skin exposure to the Sun.

Figure 15.7: Laser and directed energy beam weapons operating from satellites.

But there is another, very sinister, side to geoengineering. The large amount of metals that is being added to earth's atmosphere is making the atmosphere much more conductive. Why would that be a good thing for anybody? Well, it will be a good thing for those who wish to operate plasma discharge weapons, as they would then become much more effective. In other words, the atmosphere becomes much easier to turn into plasma, which can then be used by microwave beam generation weapons. The star wars program gave rise to many of these weapons, and they are operated from ground stations, and satellites in orbit. They can be used to destroy asteroids, such as the one that hit Russia in 2013, and they can be used to destroy factories in China. The intersection of two beams can produce explosions that can be just as powerful as nuclear explosions. So, one aim of the geoengineering crime, committed on humanity, seems to be the increased effectiveness of plasma discharge, and directed energy, weapons, as well as a way, to generate certain cloaking technology.

The diagram below shows where the microwave range, of electromagnetic waves, is, within the electromagnetic spectrum. Directed energy weapons use this range as carrier waves, and high energy beams. But some use lasers, which generate beams in the visible light range, of the spectrum.

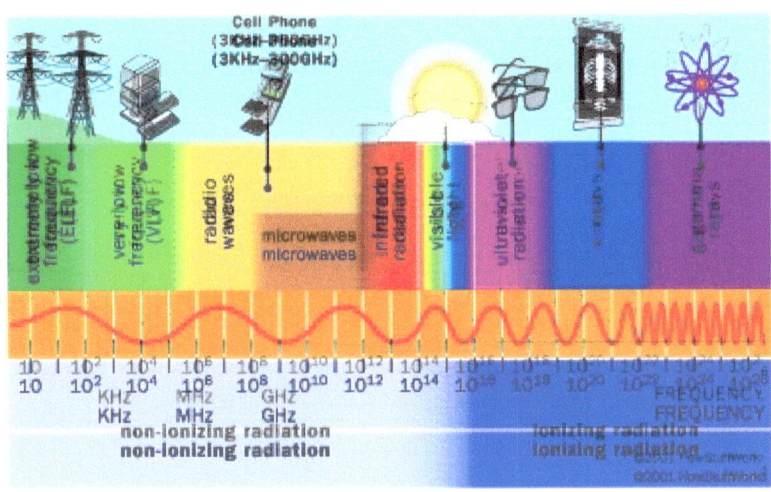

Figure 15.8: Microwaves are between the radio and infrared part of the electromagnetic spectrum and these wave frequencies are extensively used in directed energy weapons.

Another side to geoengineering is somehow even more sickening. Since we are constantly breathing in and eating these metallic nanoparticles, our bodies and brains are now filled with them and this makes it easier to read our brain patterns. The weapons, which are part of the Star Wars program are not just designed as offensive and defensive weapons, they are designed to take care of the 'powers that be' worst fear: the earth's population. They have mind control capabilities.

Figure 15.9: Asteroid streaking across the sky, in Russia, on February 15th 2013. It was destroyed by a plasma discharge or laser weapon.

Barium is sprayed in the upper atmosphere so that the Sun's ultraviolet radiation can ionize it. This barium combines with ionized aluminum to form plasma. That plasma coats the human body and can be used to track us. Every human being has a certain specific electromagnetic signature, which can be tracked by satellites, placed in orbit for that purpose. The beam constantly scans an area to find a particular person or their signature. The plasma, in the space surrounding a person, can also be used to take thoughts and images into the person's mind and a strong impulse can cause the brain to black out. In this way a person can be manipulated or even tortured and killed. However, microwaves are reflected by aluminum foil, so placing some in strategic positions around the house, or wrapping ourselves in a space blanket, should be an effective way of protecting ourselves from this invasive technology.

In conclusion, the geoengineering program seems to have four main aims: hide the objects surrounding the Sun, decrease the earth's population, make directed energy offensive and defensive weapons possible, and to possibly place the surviving population under mind control.

Chapter 16

Article 148: The purpose and effects of chemtrails

The purposeful spraying of the earth's upper atmosphere seems to have started in 1977 but it was not done on a large scale until 1997. This means that a concerted effort to change the earth's atmosphere has now been carried out for 20 years. Since whatever is added to the atmosphere eventually falls to the earth's surface, this means that lakes, rivers, oceans, soil and thus all foods are now contaminated with what has been sprayed in the upper atmosphere.

Figure 16.1. A plane sprays a deadly cocktails of chemicals, in the earth's upper atmosphere.

The chemicals contained in the aerosols sprayed by planes seem to include barium, CHAFF, strontium 90, aluminum, cadmium, chromium, nickel, radioactive thorium, mold spores, viruses, yellow fungal mycotoxins, ethylene dibromide and polymer fibers [1]. CHAFF looks like snow but it is actually fibers coated with aluminum, desiccated blood cells, plastic and paper. The fact that there have been instances where people have attempted to melt snow with fire and it has not melted, suggests that CHAFF has been dumped on the earth's surface disguised as snow.

Now the side effects associated with exposure to barium are: gastric pain, nausea, vomiting and diarrhea, followed by reduction in blood potassium levels which may lead to either high or low blood pressure, muscle weakness and kidney damage. Strontium 90 is a radioactive isotope of strontium and is produce by nuclear fission. It has a half-life of 28.8 years. It is mostly deposited in bone and bone marrow and can thus lead to bone cancer and leukemia. Early symptoms of aluminum toxicity due to exposure beyond the body's ability to excrete are flatulence, headaches, colic, dryness of the skin and mucous membranes, burning pain in the head relieved by food, heartburn and aversion to meat. Later symptoms are paralytic muscular conditions, memory loss and mental confusion. In addition, aluminum is associated with Alzheimer's disease, ALS, anemia and Parkinson's disease [1]. There is also a connection between aluminum exposure and osteoporosis [2].

Now, aluminum is the most abundant metal in the earth's crust but the human body has no use for it in any amount. However, because it can be so cheaply produced it has been added to many foods, cosmetics and medicines. It is usually added to salt, cocoa powder and baking powder as silico-aluminate in order to keep these dry. It is used as an emulsifier in processed cheese and to bleach flour, so all processed cheese and products made with bleached flour are contaminated with it. It is also used in almost all antacids and extensively used in antiperspirants. It is thus best to try and avoid those products that are known to contain it, whilst at the same time doing what we can to detoxify it, out of our bodies. Vitamin C, in the form of ascorbic acid, is known to chelate aluminum and it can thus be used to alleviate the toxic burden of aluminum on our bodies.

Figure 16.2: Vitamin C, in the form of ascorbic acid, may be helpful in detoxifying aluminum out of our bodies, as it is known to chelate aluminum.

Radioactive thorium is an alpha particle emitter and thus causes cancer. Cadmium causes respiratory tract and kidney problems. It is highly toxic even in small doses and may lead to death through renal failure. Chromium also affects the respiratory tract and may cause sinusitis, bronchitis, asthma and lung cancer as well as skin ulcers [3]. Nickel causes rashes and skin bumps, itching and dry patches of skin which resemble burns and blisters. Chronic exposure is connected with heart disease, neurological deficits, developmental deficits, in childhood, lung cancer and high blood pressure. It can also trigger inflammation leading to auto-immune disease, and is thus associated with thyroid dysfunction, fibromyalgia, lupus, diabetes and rheumatoid arthritis.

Now, the human body does need a tiny amount of nickel but we are usually getting far above those needs, since it is present in hydrogenated fats and oils, margarine, imitation butter spreads and whipping cream, any red tea, tobacco smoke, shiny stainless steel cookware and shellfish, like mussels and oysters. So it is better to try to use natural fats like real butter, cream and olive oil as far as possible. We would also in this way avoid ingesting oils made from genetically modified seeds. Genetically modified foods have a much higher herbicide load and the Roundup, the herbicide used is known to cause cancer [4].

Figure 16.3: Natural fats like butter and olive oil may be better choices than imitation and hydrogenated fats as they do not go have high amounts of nickel or contain oils form genetically modified seeds.

Mold spores are a type of fungus and can cause chronic coughing and sneezing, irritation to the eyes, mucous membranes of the nose and throat, rashes, chronic fatigue, persistent headaches, wheezing and difficulty breathing, throat irritation, chest tightness, red eyes, blurred vision, sweats, mood swings, sharp pains, abdominal pain, diarrhea, bloating, tearing, disorientation and metallic taste in the mouth. Mycotoxins are a product of mold metabolism and are associated with inflammation, allergies and infections.

Figure 16.4: Mold spores

Ethylene dibromide is a fumigant. It has been used as a post-harvest fumigant for crops in the past and is still used against termite and Japanese beetle infestation as a fumigant. It is extremely toxic to human beings and its addition to chemtrails shows that the 'powers that be' who are responsible for this spraying, think of human beings as a pest that needs to be eradicated, but slowly, so that we can spend money on pharmaceuticals, and make money for them, in the meantime.

Other reported symptoms connected with chemtrails are constipation, depression, irregular heartbeat, eye problems (nearsightedness and farsightedness, due to alterations in the interocular fluid eye pressure, and floaters), fibromyalgia, insomnia and tinnitus, which is a distant ringing in the ears or a high pitched ringing).

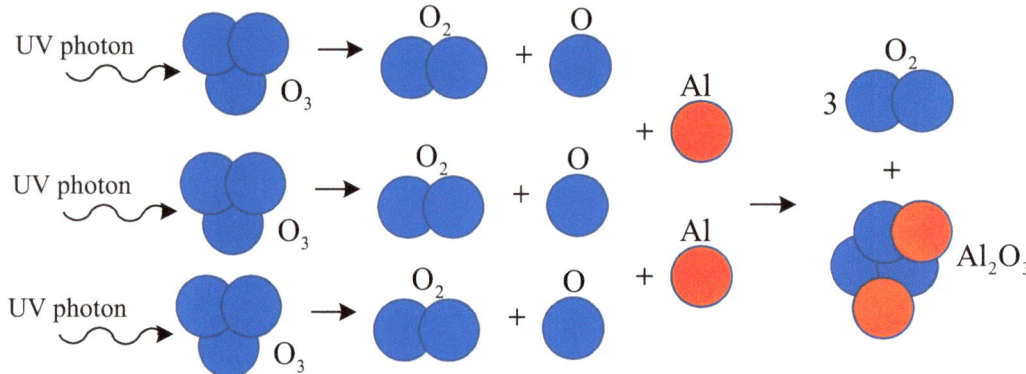

Figure 16.5. In the absence of aluminum, ozone absorbs ultraviolet (UV) radiation and ozone reforms so that it is not depleted. But, if aluminum is present in the upper atmosphere, aluminum dioxide is formed instead of ozone, so ozone is depleted.

In addition, the sometimes promoted idea that chemtrails may be useful for cooling down the planet is not at all logical. If this was the purpose, why would viruses and molds be added? Can a virus increase the amount of sunlight reflected by CHAFF? Can a virus increase the ability of chemtrail particles to act as condensation nuclei and thus produce clouds? The answer is off course, no. The only possible reason for adding something like a virus, or a fumigant, or even radioactive compounds is to make human beings ill and slowly kill them. In addition, the large amount of aluminum, in the upper atmosphere, leads to the formation of aluminum dioxide, at the expense of ozone, and in this way, depletes the ozone layer. Without the ozone layer, the planet will be sterilized by ultraviolet radiation. Therefore, continued chemtrail spraying will end all life on earth, and is thus an extinction level activity.

Why is this happening? Jesus said in John 10:10:

[10] The thief does not come except to steal, and to kill, and to destroy. I have come that they may have life, and that they may have it more abundantly.

We are living in a world run by our enemy, who is a thief. He tries to destroy us from the time we are born, and has the legal right to do so because we are born with a sinful nature and thus alienated from God. But Jesus loves us, and has died for us, so that He can forgive us, and save us. All we need to do, to access what He has paid for us, is ask Him to forgive us and take over our lives. If you have not yet done that, do it now, before it is too late.

References:

[1] https://www.globalresearch.ca/chemtrails-the-consequences-of-toxic-metals-and-chemical-aerosols-on-human-health/19047

[2] Dahl, C. et al. (2014). Do cadmium, lead, and aluminum in drinking water increase the risk of hip fractures? A NOREPOS study. Biol Trace Elem Res. 157(1):14-23. doi: 10.1007/s12011-013-9862-x.

[3] https://www.atsdr.cdc.gov/csem/csem.asp?csem=10&po=11

[4] https://www.cnn.com/2017/06/28/health/california-glyphosate-cancer-chemical-listing/index.html

Chapter 17

Article 33: Artificial weather

In this article, I explain why I think most of the weather we encounter on this planet is now artificially created and controlled. I reached this conclusion after studying the behavior of Hurricane Harvey, which devastated parts of Texas at the end of August 2017. The hurricane moved inland and then backed up thus demonstrating that it was under intelligent control. I then was directed to the website www.weatherwar101.com from which I worked out many details regarding what is happening regarding the weather patterns on our planet.

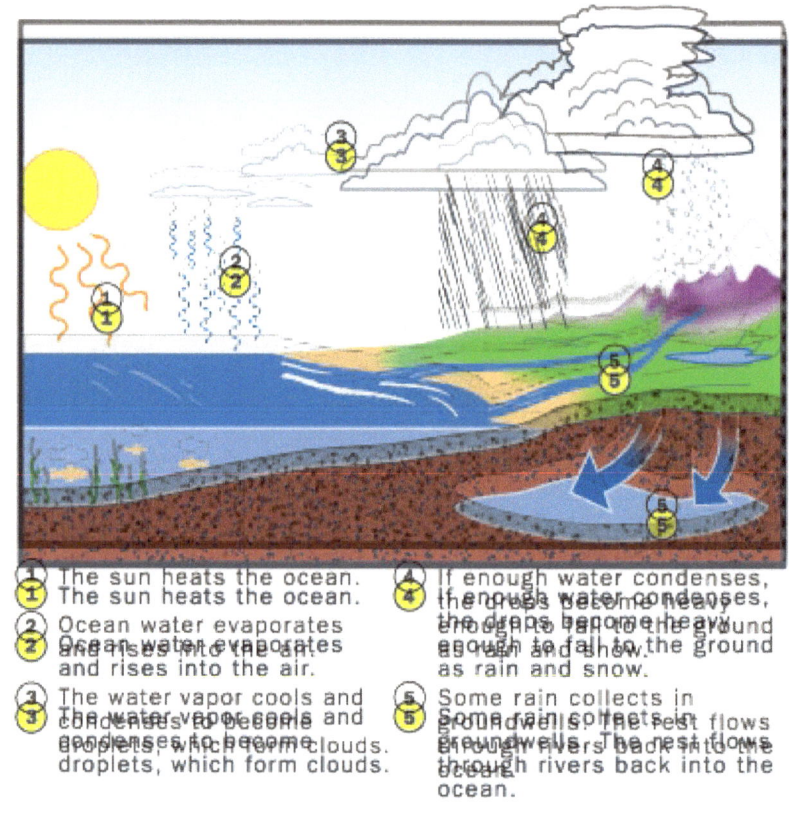

Figure 17.1: The hydrological cycle is the natural mechanism leading to rainfall.

But let me start with how rain forms. The natural way to rain formation starts with the sun heating water surfaces, such as the oceans and lakes. This then leads to evaporation and water vapor entering the atmosphere. But, in order for rain to eventually fall, this water vapor has to condense into water droplets, and in order for water droplets to form small solid particles, called Cloud Condensation Nuclei (CCNs) need to be in the atmosphere. These small particles are about the size of 0.2 micrometers or 1/100th the size of a normal sized droplet in a cloud. The water vapor condenses, or turns back to the liquid phase, on these small particles. If these solid particles are not present in the atmosphere then the water vapor would have to be cooled to subfreezing temperatures, for long periods of time, before liquid water forms. Thus, if water vapor is cooled to 8° F (-13° C), for about 6 hours then water vapor will condense into liquid droplets of water. But if subfreezing temperatures are not available then only very high pressures, would lead to the condensation of water vapor to liquid droplets. These kinds of pressures are not easily reached in a natural way. Hence, the best way to get rain is to have lots of CCNs in the atmosphere.

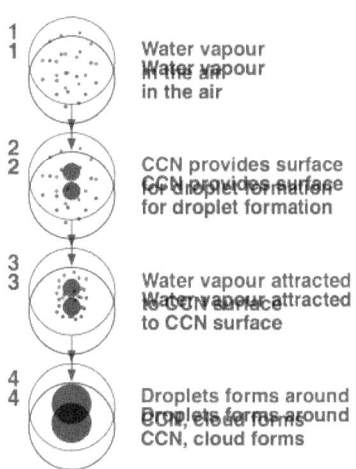

Figure 17.2: Cloud formation due to the presence of Cloud Condensation Nuclei (CCNs) in the atmosphere.

Natural CCNs are made out of dust, black carbon, from forest fires, sea salt from ocean spray, sulfate from volcanic activity and phytoplankton. Then, we also get CCNs that come from factory smokestacks and internal combustion engines; these are usually referred to as pollution as they are not natural to our atmosphere. Some CCNs are smaller than others and the ones that come from factory smoke stacks and internal combustion engines are usually much smaller, than most natural CCNs, and this is bad news for production of rain because if the CCNs are too small, the liquid droplet that condenses is too small for a raindrops to ever from.

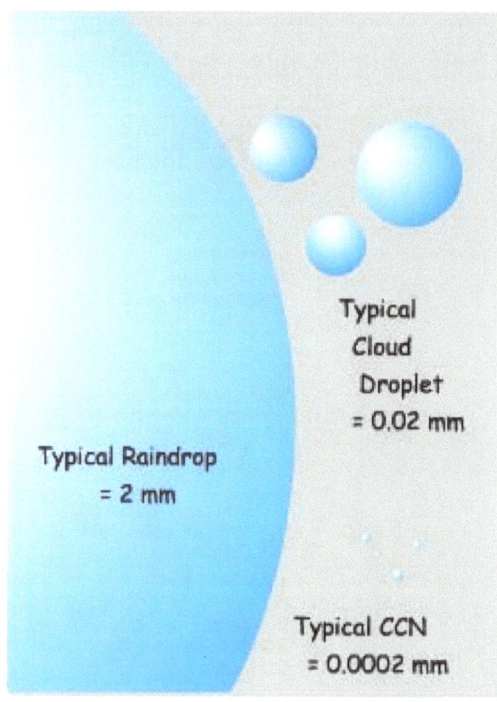

Figure 17.3. Illustration of typical sizes of CCNs, cloud droplets and raindrops.

A typical raindrop is 2 mm in diameter, whilst a typical cloud droplet is 100 times smaller, or 0.02 mm, in diameter and in order for a raindrop, which is large enough to fall to earth, to form, 1000 typical cloud drops have to collide and coalesce. So, if the drops are much smaller than 0.02 mm, it is difficult for them to collide, and coalesce, so that it becomes extremely difficult for a drop large enough to produce rain to form. This conclusion comes from a study done by Daniel Rosenfeld from the Hebrew University of Jerusalem. Daniel studied satellite images of smog streaming out of power plants, lead smelters, and oil refineries, and could see that such clouds did not lead to rain. He therefore further concluded that since this type of pollution is present, in many parts of the world, that it must be affecting rainfall on a global scale.

Figure 17.4. The burning of fossil fuels leads to the creation of CCNs, which are too small for the production of raindrops.

A report on the research done by Daniel Rosenfeld appeared on a BBC news webpage, on March 10th 2000. The report was entitled 'air pollution stops rain'. So, since for over 100 years, our world economy has been based on the internal combustion engine and the burning of fossil fuels, we should expect that this would lead to no rain, in most parts of the world. It seems that the fact that pollution resulting from the burning of fossil fuels leading to no rain would have been apparent a long time ago though, and I think it has, but instead of moving from the burning of fossil fuels, a long time ago, any technological innovation that moved away from the burning of fossil fuels has been discouraged. Why? The reason is that there are people that are got rich, and continue to get rich, from the use of coal for the production of electricity, and from the use of gas in internal combustion engines.

Figure 17.5. Resulting landscape when there is no rain.

So, instead of stopping the use of fossil fuels, another solution had to be found for the fact that the resulting pollution, would lead to no rain, and the whole planet would turn into desert like the Sahara. The solution was the production of rain artificially. For that we need a large amount of water vapor in the atmosphere. That is easily solved by letting all power stations release 50% of the steam, they produce for turning turbines and produce electricity, out into the atmosphere.

Figure 17.6: Steam coming out the cooling towers of power generation plants, which are controlled and programmed to create weather systems, lead to the creation of clouds.

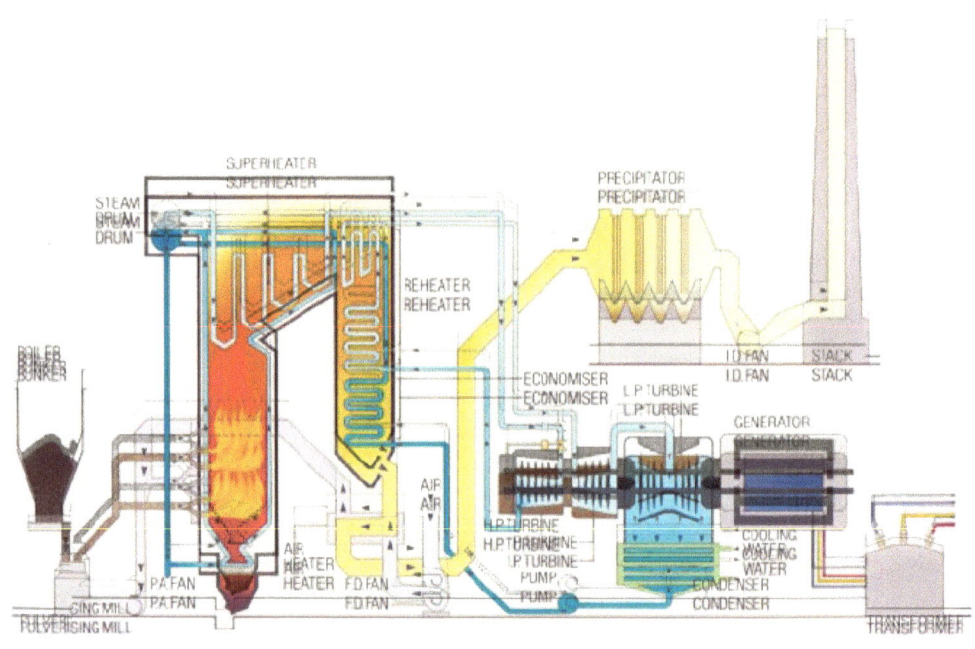

Figure 17.7: Illustration of how a coal power generation plant works; 50% of steam produced is available for weather formation and its release is programmed to occur in the correct sequence for the formation of weather systems.

Then, in order to be able to increase the severity of storms, we need a way to compress spinning cloud formations, thus causing them to spin faster due to the principle of conservation of angular momentum. This is accomplished through the production of acoustic waves in air. This only requires an oscillation of air masses, which is accomplished by radar beams, hence the need for radar installations in all land masses, where weather patterns are artificially produced. Then, in order to make the weather patterns or clouds easy to steer, it is important to make clouds electrically conductive. This is accomplished by

98

filling the atmosphere with CCNs, made out of heavy metals. This is achieved through the heavy use of aluminum oxides in chemtrails. And thus we have the production of artificial weather.

Now, the fact that we now have large numbers of Stellar remnants close to the Sun and destabilizing it as well as heating our planet from its core has made the weather problem more severe and has also led to the increase in the use of chemtrails, which are poisoning the planet and destroying the ozone layer. This may lead to a tolerable temperature, on the surface, and continued rain, for a while longer, but the UV C radiation will eventually kill every living organism on this planet. So solutions that preserve the planet and life on it should be sought.

Figure 17.8: Chemtrails besides many purposes including the hiding of the Stellar Remnants gathered in the Sun's Corona, chemtrails are also used for the purpose of creating artificial weather.

Also, it seems that now, the people who have taken over the production of artificial weather have also declared war on certain countries, including the United States. If any government would like to fight back and survive, it may be necessary to take over radar installations, or build microwave beam generators. It seems that beams are aimed in such a way as to compress cloud systems, and lead to a faster rotation, and thus increase the severity of the storm. This could possibly be reversed by aiming radar or microwave beams at the eye of the storm. This should cause it to slow its rotation and possibly dissipate.

In conclusion, artificial weather seems to be widespread throughout the whole planet but now that it is been used as a weapon against certain countries, it becomes necessary to build countering devices and mechanisms. If any government on this planet wishes to protect their own people and survive as a nation, they should fight back and work on dissipating destructive weather systems, and also work on eliminating the pollution problem by moving away from burning fossil fuels.

Chapter 18

Article 137: Noctilucent clouds, rocket launches and chemtrails: what are they hiding?

Noctilucent clouds formed over Japan, after a rocket launch, on January 18th 2018. This is not the first time that noctilucent clouds were seen over Japan, after a rocket launch. These clouds were also seen over Japan, after another Japanese rocket launch, in 2015. On January 18th 2018, the effect seemed more pronounced though, with the clouds seeming to form as the rocket streaked across the sky. They were also somehow similar to the phenomena observed during the SpaceX launch, on December 22nd 2017.

Figure 18.1. Noctilucent clouds form behind a Japanese rocket as it streaks across the sky.

Figure 18.2. Noctilucent clouds observed in the sky over Japan, on November 24th 2015, after a rocket launch.

Figure 18.3. The Space X falcon 9 launch on December 22nd 2018 produced a strange effect in the sky which is similar to what was observed in association with the Japanese rocket launch of January 18th 2018.

Noctilucent clouds, are luminescent clouds which can only be observed after the Sun has sunk below the horizon. They form at very high altitudes, at around 80 km or 50 miles above sea level, in layer of the atmosphere called the mesosphere. They seem to be made of tiny ice crystals and possibly are formed around micrometer dust, or directly from water vapor directly, because of the very low temperatures which occur at that altitude. It is possible for water vapor to sublimate directly into ice, at the low temperatures and pressures, of the mesosphere, but once ice crystals form, how do they stay up there since they would be so much denser than the air around them?

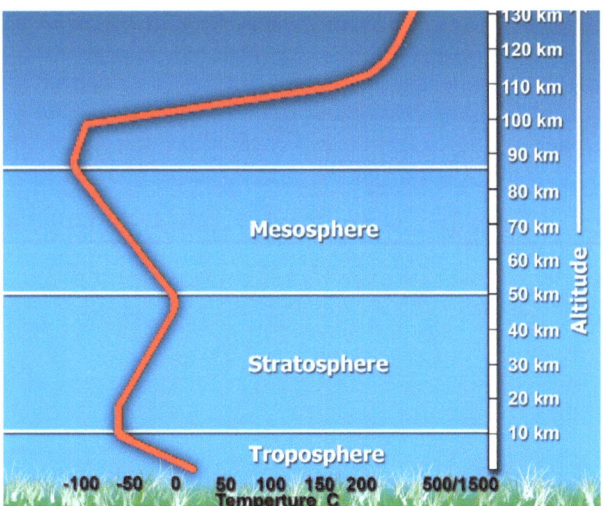

Figure 18.4. Noctilucent clouds form in the Mesosphere at an altitude of between 76 km and 85km. At that altitude the air temperature varies between -75 °C and -100 °C.

Now, clouds in the earth's atmosphere which form at normal altitudes of between 30 000 feet and 45 000 feet, which corresponds to between 9 km and 14 km, or 6 and 9 miles, are also heavier than air and stay suspended in the atmosphere due to an electric effect as explained in article 122: Electric weather: why is it getting more severe? Water molecules are naturally dipoles and move along electric fields, and thus electric fields arising from induced currents, in the earth's ionosphere, cause water molecules to arise above low pressure areas as illustrated in the figure below.

Figure 18.5: Currents in the ionosphere induce currents in the earth's lower atmosphere and on the earth's surface, and create low and high pressure regions, which are centers of circulation due to the direction of the electric field at that point. Low pressures lead to water molecules spiraling upward and thus the formation of clouds, whilst high pressure areas draw water molecules downwards. Stellar Cores increase ionization in the ionosphere and therefore create increased circulation of air.

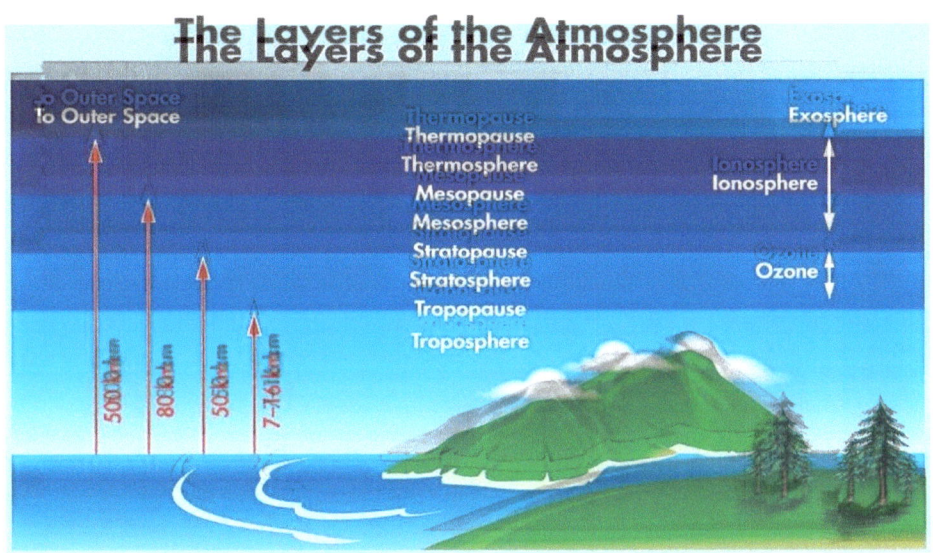

Figure 18.6: The mesosphere, where noctilucent clouds form, is part of the lower ionosphere.

Now, the mesosphere, where noctilucent clouds form, is a part of the ionosphere, so these may be forming as a result of increased ionization, in that part of the earth's atmosphere, where solar wind particles, solar radiation create ionized atmospheric particles and thus electric currents. The fact that these clouds usually only form close to the poles, also suggests that these are associated to electric effects, in the ionosphere, as the solar wind particles reach the ionosphere, more easily over the poles. However, these clouds have been occurring at lower latitudes more recently, which may indicate increased ionization and increased bombardment of the earth's ionosphere by particles and radiation from space.

Figure 18.7: Drawing of the Krakatoa volcano, which erupted in 1883, and destroyed the island. The eruption created 120 feet high tsunamis, which killed an estimated 36 000 people on other nearby islands. Noctilucent clouds appeared 2 years later and were initially connected to the eruptions but have continued to appear since then.

Noctilucent clouds were not observed, with any frequency, until 1885, but may have been observed for the first time in 1850. They were first believed to be connected with the eruption of the Krakatoa, which erupted 2 years before the clouds started appearing. However, although volcanoes usually place large amounts of dust, in the atmosphere, which provide cloud nucleation centers, around which water vapor can condense and form drops of water, and thus lead to an increase in the rate of cloud formation, in the earth's atmosphere, these clouds form normal cloud formation altitude, which is between 9 km and 14 km, or 6 and 9 miles. In addition, there had cataclysmic volcanic eruptions before, and none had led to the observance of extreme high altitude luminescent clouds. And the noctilucent clouds have continued appearing and seem to be getting more prevalent and although initially only believed to occur close to the poles, they have started appearing at lower latitudes.

Now, the first documented report of a rejuvenated Stellar Core being observed, from earth, comes from Fatima in 1917, and it is thus possible that these objects may have started arriving by 1885, which may have led to an increased ionization of the earth's ionosphere, at that time, in addition to the arrival of debris dust from these objects, which may have provided nucleation centers, in the upper atmosphere.

Since noctilucent have been occurring since the last half of the 19th century but did not seem to have been observed before then, they point to something having changed in the Solar System and having caused their appearance. We know this is most likely to be the appearance of the Stellar Core system close to the Sun.

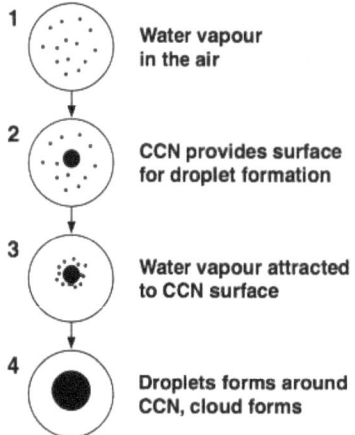

Figure 18.8. Cloud formation due to the presence of Cloud Condensation Nuclei (CCNs) in the atmosphere.

Cloud formation usually requires nucleation centers, also called cloud condensation nuclei (CCN) and these may have been provided by dust debris coming from the Stellar Core system. Cloud formation also requires ionization, in the ionosphere, and thus the appearance of the clouds, indicate that increased ionization may have something to do with the appearance of noctilucent clouds. The fact that they were initially believed to only occur at very high altitudes, or close to the poles, shows that this ionization and possibly the debris dust has increased, which again points to the Stellar Core system having an increased influence.

Figure 18.9. The strange clouds and colors, appearing in the sky, at the same time as, the Japanese rocket launch. It is possible that the rocket launch and production of artificial noctilucent clouds was to hide the appearance, of a blue light emitting Stellar Core followed by a pink and blue tail, and that this object was actually out in space and much further than the rocket.

Rocket launches have been taking place since the 1960s, and yet they have never before been associated with the formation of luminescent clouds, this seems to be a recent development, and points to something having changed in the earth's atmosphere. This could be due to the same ionization and debris brought in by the Stellar Core system, which is likely to have led to the appearance of the clouds in the first place. It could also be due to the spraying of dust or artificial nucleation centers in the mesosphere, by the rockets, which would then lead to the appearance of these types of clouds forming behind the rocket and prevailing in the sky afterwards due to the natural additional ionization caused by the Stellar Cores. In this case, the same reason for the spraying of aerosols or chemtrailing is done lower down in the earth's atmosphere: to hide the presence of the Planet X, or Stellar Core, system, from the earth's population.

Stellar Cores are often surrounded by clouds of ionized material, which may emit light. This is as a result of the fact that they have aged, by going through the red giant, and white dwarf, phases, and have lost most of their layers of ionizing material, with some of that material remaining around the core and forming a cloud. Once an object has obtained enough electric potential from the Sun, it may be able to ionize this cloud to the point that it emits visible light, and thus be visible in the sky, from the earth's surface. Creating artificial noctilucent clouds may be a way of hiding these clouds and objects, which would be visible in earth's skies, from the earth's population.

Figure 18.10: Photograph taken on January 20th 2018 of chemtrail lines, across the sky that could only have been created by 4 chemtrail planes, flying in parallel formation. This is a degree of chemtrailing effort that is not often observed.

Figure 18.11: Photographs of the sky in Pennsylvania, USA, on January 20th 2018, showing that a huge effort is being placed into covering the sky with chemtrail clouds.

The degree of chemtrailing, in the earth's atmosphere, as indicated by figures 10 and 11, seems to have reached a new level showing a degree of desperation on the part of the 'powers that be' to hide the Stellar Cores, which seem to be more numerous and brighter and thus more difficult to hide now than ever before. The strange cloud formations, now occurring as a result of rocket launches, are likely to be an additional effort being carried out with the same end in mind.

In conclusion, the appearance of noctilucent clouds in the second half of the 19th century seems to be due the appearance of the first members of the Planet X, or Stellar Core, System, in the Solar System. The increased incidence, and their appearance, at lower latitudes, is most likely due to the increased numbers, and effect, these objects are having on planet earth. Increased effort in chemtrailing the atmosphere indicates that these objects have become more difficult to hide. The recent appearance of noctilucent clouds, in association with rocket launches, is likely to be due to the purposeful spraying of aerosols, from these rockets, of the mesosphere, and thus producing extreme high altitude chemtrail clouds, in order to again hide the presence of these objects from the earth's population.

The End for Now!

www.ingramcontent.com/pod-product-compliance
Lightning Source LLC
Chambersburg PA
CBHW051913210526
45473CB00006B/2001